P9-DBM-274

MATH MATTERS
for Adults

Decimals and Percents

Author
Karen Lassiter
Austin Community College
Austin, Texas

Consultants
Connie Eichhorn
Omaha Public Schools
Omaha, Nebraska

M. Gail Joiner Ward
Swainsboro Technical Institute
Swainsboro, Georgia

STECK-VAUGHN
C O M P A N Y
A Subsidiary of National Education Corporation

About the Author

Dr. Karen Lassiter is currently a mathematics instructor at Austin Community College. She is a former Senior Math Editor for Steck-Vaughn Company, and has done extensive work with standardized test preparation. Dr. Lassiter holds a Ph.D. in Educational Research, Testing, and Instructional Design and a bachelor's degree in mathematics and science education from Florida State University.

About the Consultants

Connie Eichhorn is a supervisor of adult education programs in the Omaha Public Schools. A former mathematics teacher and ABE/GED instructor, she earned an undergraduate degree in math at Iowa State University and is completing a doctoral program in adult education at the University of Nebraska. She conducts workshops on math instruction for teachers of adult basic education.

M. Gail Joiner Ward is presently working at Swainsboro Technical Institute as an adult education instructor. Previously she worked with adult learners in a multimedia lab through the School of Education at Georgia Southern University. She has bachelor's and master's degrees in art and early childhood education from Valdosta State College and Georgia Southern University.

Staff Credits

Executive Editor:	Ellen Lehrburger
Design Manager:	Pamela Heaney
Illustration Credits:	David Griffin, Alan Klemp, Mike Krone, Kristian Gallagher
Photo Credits:	Cover: (inset) Cooke Photographics, (background) © Park Street; P.9 © Park Street; p.31 © Tony Freeman/PhotoEdit; p.47 © Bob Daemmrich/Stock Boston; p.73 © Park Street; p.101 © Park Street; p.141 © Focus on Sports.
Cover Design:	Pamela Heaney

ISBN 0-8114-3652-7

Copyright © 1993 Steck-Vaughn Company

All rights reserved. No part of the material protected by this copyright may be reproduced or utilized in any form or by any means, electronic or mechanical, including photocopying, recording, or by any information storage and retrieval system, without permission in writing from the copyright owner. Requests for permission to make copies of any part of the work should be mailed to: Copyright Permissions, Steck-Vaughn Company, P.O. Box 26015, Austin, Texas 78755.

Printed in the United States of America.
1 2 3 4 5 6 7 8 9 CK 98 97 96 95 94 93

Contents

Unit  1

The Meaning of Decimals

Unit 2

Adding and Subtracting Decimals

Unit 3

Multiplying and Dividing Decimals

Unit 4

Ratios, Proportions, and Percents

Unit 5

USING PERCENTS

Unit 6

PUTTING YOUR SKILLS TO WORK

TO THE LEARNER

The four books in the Steck-Vaughn series *Math Matters for Adults* are *Whole Numbers; Fractions; Decimals and Percents;* and *Measurement, Geometry, and Algebra.* They are written to help you understand and practice arithmetic skills, real-life applications, and problem-solving techniques.

This book contains features which will make it easier for you to work with decimals and to apply them to your daily life.

A Skills Inventory appears at the beginning and end of the book.
• The first test shows you how much you already know.
• The final test can show you how much you have learned.

Each unit has several Mixed Reviews and a Unit Review.
• The Mixed Reviews give you a chance to practice the skills you have learned.
• The Unit Review helps you decide if you have mastered those skills.

There is also a glossary at the end of the book.
• Turn to the glossary to find the meanings of words that are new to you.
• Use the definitions and examples to help strengthen your understanding of terms used in mathematics.

The book contains answers and explanations for the problems.
• The answers let you check your work.
• The explanations take you through the steps used to solve the problems.

Decimals and Percents Skills Inventory

Write as a decimal.

1. six hundredths   2. two and one tenth   3. two dollars   4. four cents

Change to a decimal.

5. $\frac{1}{4} =$ 6. $\frac{3}{10} =$ 7. $\frac{2}{3} =$ 8. $5\frac{1}{5} =$

Change to a mixed number or fraction. Reduce if possible.

9. .9 =   10. .01 =   11. 6.25 =   12. 12.87 =

Compare. Write <, >, or = in each box.

13. .1 ☐ .100   14. 1.3 ☐ 1.03   15. .72 ☐ .70   16. 5.6 ☐ 6.5

Round to the nearest whole number.

17. 1.3 _____   18. .9 _____   19. 2.51 _____   20. 7.2 _____

Round to the nearest tenth.

21. 4.28 _____   22. .73 _____   23. .932 _____   24. 17.85 _____

Round to the nearest hundredth.

25. .266 _____   26. .019 _____   27. 1.199 _____   28. 5.4758 _____

Add, subtract, multiply, or divide. Round division answers to the nearest hundredth.

29.	30.	31.	32.	33.
$9.27 + 4.86	27.39 + 11.7	$5.00 − 1.49	95.3 − 1.052	7 − 3.68

34.	35.	36.	37.	38.
$9.25 × 8	$ .97 × 100	1.45 × 8.2	7.859 × .06	.2 × .4

39. $5\overline{)\$67.35}$ **40.** $12\overline{)\$9.98}$ **41.** $10\overline{)\$.90}$ **42.** $.7\overline{).15}$ **43.** $2.8\overline{)4.37}$

Write a ratio.

44. 5 hits and 6 misses   **45.** 10 chances in 100   **46.** two wins and two losses

Solve each proportion.

47. $\dfrac{4}{8} = \dfrac{n}{4}$ **48.** $\dfrac{2}{3} = \dfrac{4}{n}$ **49.** $\dfrac{n}{5} = \dfrac{6}{10}$ **50.** $\dfrac{1}{n} = \dfrac{5}{10}$

Write a percent using the percent sign.

51. one percent =   **52.** two and three tenths percent =

Change to a decimal or a percent.

53. $50\% =$ **54.** $205\% =$ **55.** $7\frac{1}{4}\% =$ **56.** $33\frac{1}{3}\% =$

57. $.08 =$ **58.** $.3 =$ **59.** $.75 =$ **60.** $2 =$

Change to a whole number or a fraction. Reduce if necessary.

61. $25\% =$ **62.** $7\% =$ **63.** $300\% =$ **64.** $2\frac{1}{2}\% =$

Change to a percent.

65. $\dfrac{3}{4} =$ **66.** $\dfrac{3}{100} =$ **67.** $\dfrac{1}{6} =$ **68.** $\dfrac{3}{5} =$

Compare. Use <, >, or = sign.

69.

.2 ☐ 20%

70.

70% ☐ 7

71.

$\frac{1}{2}$ ☐ 2%

72.

$33\frac{1}{3}\%$ ☐ $\frac{1}{3}$

Solve.

73.
What is 5% of 90?

74.
What is $33\frac{1}{3}\%$ of 120?

75.
What is 110% of 10?

76.
3 is 10% of what number?

77.
50 is 200% of what number?

78.
15 is $1\frac{1}{2}\%$ of what number?

79.
13 is what percent of 78?

80.
10 is what percent of 5?

**Below is a list of the problems in this Skills Inventory and the pages
on which the skills are taught. If you missed any problems, turn to
the pages listed and practice the skills. Then correct the problems you
missed in the Skills Inventory.**

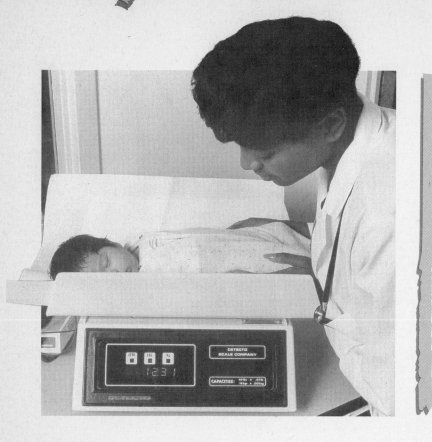

Decimals are fractions in a different form. The fraction $\frac{1}{2}$ means the same as the decimal .50. You know this because half a dollar is fifty cents or $ .50. Money and weights and measures are the most common uses of decimals.

In this unit, you will learn how to write decimals, change fractions to decimals, change decimals to fractions, round decimals, and use decimals to solve problems.

Getting Ready

You should be familiar with the skills on this page and the next before you begin this unit. To check your answers, turn to page 159.

▶ You can use a place value chart to help you write whole numbers.

305	three hundred five
62	sixty-two
1,324	one thousand, three hundred twenty-four

Write the following numbers in the place value chart on the right. Then write each number in words on the lines below.

thousands	hundreds	tens	ones
	3	0	5
		6	2
1,	3	2	4
	5	5	7

1. 557 _____ five hundred fifty-seven

2. 99 _____

3. 8 _____

4. 2,040 _____

> To change a fraction to higher terms, multiply the numerator and the denominator by the same number.

Change the following fractions to higher terms.

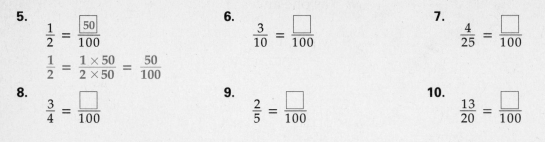

5.

$$\frac{1}{2} = \frac{\boxed{50}}{100}$$

$$\frac{1}{2} = \frac{1 \times 50}{2 \times 50} = \frac{50}{100}$$

6.

$$\frac{3}{10} = \frac{\square}{100}$$

7.

$$\frac{4}{25} = \frac{\square}{100}$$

8.

$$\frac{3}{4} = \frac{\square}{100}$$

9.

$$\frac{2}{5} = \frac{\square}{100}$$

10.

$$\frac{13}{20} = \frac{\square}{100}$$

For review, see pages 19–20 in **Math Matters for Adults, Fractions.**

> To reduce a fraction to lowest terms, divide the numerator and the denominator by the same number.

Change the following fractions to lowest terms.

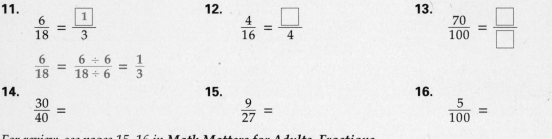

11.

$$\frac{6}{18} = \frac{\boxed{1}}{3}$$

$$\frac{6}{18} = \frac{6 \div 6}{18 \div 6} = \frac{1}{3}$$

12.

$$\frac{4}{16} = \frac{\square}{4}$$

13.

$$\frac{70}{100} = \frac{\square}{\square}$$

14.

$$\frac{30}{40} =$$

15.

$$\frac{9}{27} =$$

16.

$$\frac{5}{100} =$$

For review, see pages 15–16 in **Math Matters for Adults, Fractions.**

> A mixed number has a whole number and a fraction. To change a mixed number to lower or higher terms, change only the fraction.

Change to lowest terms.

17.

$$4\frac{6}{8} = 4\frac{6 \div 2}{8 \div 2} = 4\frac{3}{4}$$

18.

$$3\frac{3}{9} =$$

19.

$$1\frac{5}{25} =$$

20.

$$7\frac{10}{100} =$$

21.

$$18\frac{12}{50} =$$

22.

$$13\frac{20}{100} =$$

Change to higher terms.

23.

$$10\frac{3}{8} = 10\frac{\boxed{9}}{24}$$

$$10\frac{3}{8} = 10\frac{3 \times 3}{8 \times 3} = 10\frac{9}{24}$$

24.

$$2\frac{1}{4} = 2\frac{\square}{16}$$

25.

$$32\frac{2}{5} = 32\frac{\square}{10}$$

For review, see pages 27–28 in **Math Matters for Adults, Fractions.**

Writing Decimals

Decimals are another way of writing fractions. The shaded part of each square below can be written as a fraction and a decimal.

A decimal point separates the whole number and decimal part. A decimal point is read as *and*. For example, you read 1.5 as one and five tenths.

A whole number can be written as a decimal by placing a decimal point followed by a zero to the right of the number. For example, the whole number 1 can be written as the decimal 1.0.

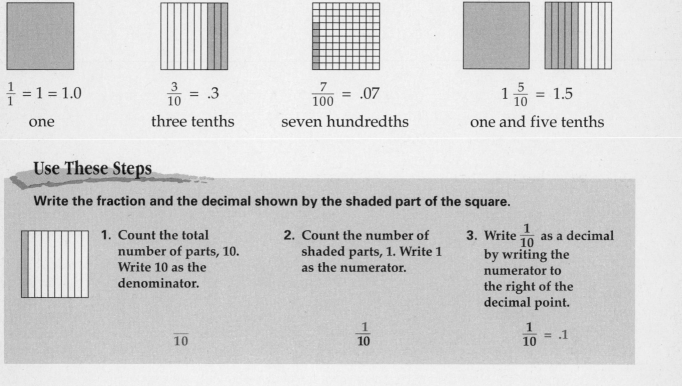

$\frac{1}{1} = 1 = 1.0$

one

$\frac{3}{10} = .3$

three tenths

$\frac{7}{100} = .07$

seven hundredths

$1\frac{5}{10} = 1.5$

one and five tenths

Use These Steps

Write the fraction and the decimal shown by the shaded part of the square.

1. Count the total number of parts, 10. Write 10 as the denominator.

2. Count the number of shaded parts, 1. Write 1 as the numerator.

3. Write $\frac{1}{10}$ as a decimal by writing the numerator to the right of the decimal point.

$\overline{10}$

$\frac{1}{10}$

$\frac{1}{10} = .1$

Write the fraction and the decimal shown by the shaded part of each square.

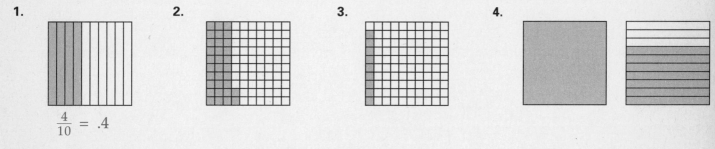

1.

$\frac{4}{10} = .4$

2.

3.

4.

Write a decimal for each fraction or mixed number.

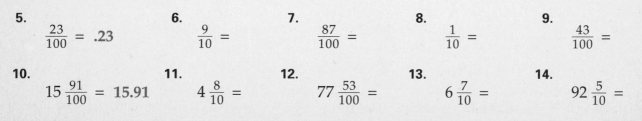

5. $\frac{23}{100} = .23$

6. $\frac{9}{10} =$

7. $\frac{87}{100} =$

8. $\frac{1}{10} =$

9. $\frac{43}{100} =$

10. $15\frac{91}{100} = 15.91$

11. $4\frac{8}{10} =$

12. $77\frac{53}{100} =$

13. $6\frac{7}{10} =$

14. $92\frac{5}{10} =$

Changing Fractions to Decimals

It is easy to write fractions such as $\frac{1}{2}$, $\frac{1}{4}$, or $\frac{1}{5}$ as decimals. First change the fraction to a fraction in higher terms with 10 or 100 as the denominator. Then write the numerator to the right of the decimal point.

Use These Steps

Write $\frac{1}{5}$ as a decimal.

1. Multiply the numerator and denominator by 2.

2. Write the numerator, 2, to the right of the decimal point.

$$\frac{1}{5} = \frac{1 \times 2}{5 \times 2} = \frac{2}{10}$$

.2

Change each fraction or mixed number to a decimal.

1. $\frac{3}{4} = \frac{3 \times 25}{4 \times 25} = \frac{75}{100} = .75$

2. $\frac{1}{2} =$

3. $\frac{11}{25} =$

4. $\frac{4}{5} =$

5. $\frac{3}{20} =$

6. $\frac{9}{25} =$

7. $\frac{3}{5} =$

8. $\frac{7}{20} =$

9. $\frac{1}{4} =$

10. $\frac{21}{25} =$

11. $\frac{29}{50} =$

12. $\frac{37}{50} =$

13. $7\frac{3}{25} = 7\frac{3 \times 4}{25 \times 4} = 7\frac{12}{100} = 7.12$

14. $36\frac{7}{25} =$

15. $5\frac{1}{2} =$

16. $1\frac{2}{5} =$

17. $25\frac{3}{4} =$

18. $33\frac{3}{10} =$

19. $104\frac{4}{5} =$

20. $90\frac{1}{20} =$

Writing Tenths

A decimal with one digit to the right of the decimal point means tenths. For example, two tenths of a mile is written as $\frac{2}{10}$ or .2.

|←————1 mile————→|

two tenths = $\underset{10}{\underline{2}}$ ←— parts shaded = .2
 ←— parts in all

Use These Steps

Write five tenths as a fraction. Then change to a decimal.

1. Write the fraction.

$\frac{5}{10}$

2. Write the numerator to the right of the decimal point.

.5

Write the number as a fraction or mixed number. Then write it as a decimal.

1. one tenth
$\frac{1}{10}$ = .1

2. six tenths

3. seven tenths

4. four tenths

5. nine tenths

6. two tenths

7. seven and five tenths
$7\frac{5}{10}$ = 7.5

8. ten and two tenths

9. ninety-six and six tenths

10. nine and nine tenths

11. thirty and no tenths
$30\frac{0}{10}$ = 30.0

12. fifteen and no tenths

13. forty-two and nine tenths

14. eighty-six and three tenths

15. sixty-seven and no tenths

16. seventy and five tenths

17. six and no tenths

18. ninety-nine and no tenths

19. fifty-three and six tenths

20. twenty-one and two tenths

21. eighty and four tenths

22. thirty-two and nine tenths

Writing Hundredths

A decimal with two digits to the right of the decimal point means hundredths.

Rainfall is often measured in hundredths of an inch. Fifteen hundredths of an inch of rain is written $\frac{15}{100}$ or .15.

$$\text{fifteen hundredths} = \frac{15}{100} = .15$$

Two hundredths of an inch of rain is written $\frac{2}{100}$ or .02. Since 2 only uses one place, insert a zero next to the decimal point to fill the other place.

$$\text{two hundredths} = \frac{2}{100} = .02$$

└──insert a zero

Use These Steps

Write six hundredths as a fraction. Then change to a decimal.

1. Write the fraction.

$$\frac{6}{100}$$

2. Write the numerator to the right of the decimal point. Since the numerator has one digit, insert a zero between the numerator and the decimal point.

.06

insert a zero

Write the number as a fraction or a mixed number. Then write it as a decimal.

1. seventeen hundredths

$\frac{17}{100} = .17$

2. twenty-six hundredths

3. eleven hundredths

4. fifty hundredths

5. ten hundredths

6. seventy hundredths

7. three hundredths

$\frac{3}{100} = .03$

8. one hundredth

9. eight hundredths

10. ten and twelve hundredths

$10\frac{12}{100} = 10.12$

11. thirty and one hundredth

12. six and no hundredths

13. ninety-nine and no hundredths

14. fifty-three and six hundredths

15. twenty-one and ten hundredths

16. eighty and fourteen hundredths

17. thirty-two and fifty-nine hundredths

Real-Life Application   At Work

Decimals show parts of measures such as money, time, and weight. Sometimes you need to change part of something to a decimal. For example, thirty minutes is half an hour or .5 hours as a decimal.

Example On his timecard, Fernando needs to write in decimals the number of hours that he works. Sunday he worked $7\frac{1}{4}$ hours. What will he write on his timecard?

Change the mixed number to a decimal with two digits to the right of the decimal point.

$$7\frac{1}{4} = 7\frac{1 \times 25}{4 \times 25} = 7\frac{25}{100} = 7.25$$

Fernando will write 7.25 on his timecard.

Sunday	7	25
Monday		
Tuesday		
Wednesday		
Thursday		
Friday		
Saturday		

Change the hours Fernando worked to decimals. Complete his timecard for the week.

1. Monday Fernando worked for 4 hours.

   Answer_____

2. Tuesday he worked for $3\frac{1}{2}$ hours.

   Answer_____

3. Wednesday was his day off.

   Answer_____

4. Thursday he worked for $6\frac{1}{4}$ hours.

   Answer_____

5. Friday he worked for $8\frac{3}{4}$ hours.

   Answer_____

6. Saturday he put in $2\frac{1}{2}$ hours.

   Answer_____

Changing Decimals to Fractions

When you change a decimal to a fraction with a denominator of 10 or 100, write the digits to the right of the decimal point as the numerator of the fraction.

Notice that the number of places to the right of the decimal point is the same as the number of zeros in the denominator of the fraction.

Use These Steps

Change .08 to a fraction.

1. Write the digits to the right of the decimal point as the numerator. Omit the zero.

2. There are two places to the right of the decimal point, so the denominator needs two zeros. Write 100. Reduce.

$$.08 = \frac{8}{}$$

$$\frac{8}{100} = \frac{8 \div 4}{100 \div 4} = \frac{2}{25}$$

Change each decimal to a fraction. Reduce if possible.

1.
$$.7 = \frac{7}{10}$$

2.
$$.6 =$$

3.
$$.9 =$$

4.
$$.1 =$$

5.
$$1.3 = 1\frac{3}{10}$$

6.
$$6.5 =$$

7.
$$10.4 =$$

8.
$$8.7 =$$

9.
$$.57 =$$

10.
$$.61 =$$

11.
$$.92 =$$

12.
$$.60 =$$

13.
$$3.44 =$$

14.
$$12.99 =$$

15.
$$15.13 =$$

16.
$$29.25 =$$

17.
$$.05 =$$

18.
$$.09 =$$

19.
$$.01 =$$

20.
$$.07 =$$

21.
$$11.03 =$$

22.
$$29.06 =$$

23.
$$37.02 =$$

24.
$$75.01 =$$

25.
$$1.83 =$$

26.
$$94.76 =$$

27.
$$.45 =$$

28.
$$25.21 =$$

29.
$$43.66 =$$

30.
$$6.1 =$$

31.
$$7.2 =$$

32.
$$9.93 =$$

Money and Decimals

Amounts of money are almost always written as decimals with two places after the decimal point. For example, two dollars and fifty-nine cents is written as $2.59.

Use These Steps

Write six dollars and three cents as a decimal.

1. Write the dollar sign. Then write the whole dollar amount followed by a decimal point.

   $6.

2. Write three cents to the right of the decimal point. Insert a zero between the decimal point and the 3.

   $6.03

   ↖ insert a zero

Write each amount as a decimal.

1. seventy-two cents
   $ .72

2. nine cents

3. fifty cents

4. eighty cents

5. one dollar and one cent

6. five dollars and seven cents

7. ten dollars and forty cents

8. nineteen dollars and sixty-six cents

9. forty-seven dollars

10. twenty dollars

Write each amount in words.

11. $ .48 _forty-eight cents_____

12. $ .92 _____

13. $ .30 _____

14. $ .20 _____

15. $ .01 _____

16. $ .05 _____

17. $4.09 _____

18. $11.02 _____

19. $7.60 _____

20. $20.10 _____

21. $32.49 _____

22. $75.00 _____

Mixed Review

Write the shaded part of each square as a whole number, fraction, or mixed number. Then change to a decimal.

1.

2.

3.

Change each fraction or mixed number to a decimal.

4. $\frac{1}{2} =$

5. $1\frac{3}{5} =$

6. $27\frac{19}{20} =$

7. $2\frac{7}{10} =$

8. $\frac{4}{5} =$

9. $32\frac{7}{10} =$

10. $12\frac{8}{25} =$

11. $4\frac{87}{100} =$

12. $63\frac{11}{20} =$

13. $1\frac{9}{100} =$

14. $72\frac{3}{25} =$

15. $4\frac{9}{10} =$

Write a fraction or mixed number. Then change to a decimal.

16. six tenths

17. six hundredths

18. one and no hundredths

19. nine and one tenth

20. forty hundredths

21. fourteen and two tenths

Change each decimal to a mixed number or fraction. Reduce if possible.

22. .07 =

23. 4.5 =

24. 89.99 =

25. .33 =

26. 22.4 =

27. 69.12 =

Write each amount as a decimal.

28. two dollars

29. ten cents

30. four dollars and six cents

Write each amount in words.

31. $ .40 _____

32. $106.22 _____

Place Value

You have worked with decimals with one place (tenths) and two places (hundredths) to the right of the decimal point. The place value chart below shows decimals with three places (thousandths) and four places (ten thousandths).

six tenths

fifty-four hundredths

two hundred one thousandths

five ten thousandths

tens	ones	.	tenths	hundredths	thousandths	ten thousandths
		.	6			
		.	5	4		
		.	2	0	1	
		.	0	0	0	5

Use These Steps

Write the decimal 1.306 in words.

1. Write the whole number in words. Write the word *and* for the decimal point.

   one and

2. Write the decimal part. Then write the place name of the last digit.

   one and **three hundred six thousandths**

Put each number in the chart to the right. Then write each decimal in words.

1. 60.3 — sixty and three tenths

2. 19.7 _____

3. 1.07 _____

4. 100.11 _____

5. .89 _____

6. 2.49 _____

7. 11.299 _____

8. 99.035 _____

9. .0016 _____

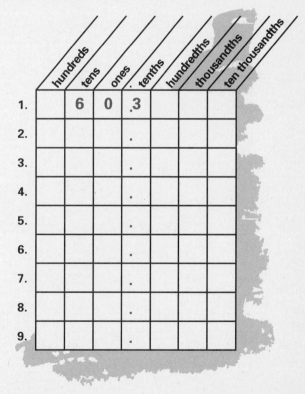

	hundreds	tens	ones	.	tenths	hundredths	thousandths	ten thousandths
1.		6	0	.	3			
2.				.				
3.				.				
4.				.				
5.				.				
6.				.				
7.				.				
8.				.				
9.				.				

Zeros and Decimals

You can add zeros to the right of a decimal without changing its value.

$$\$10 = \$10.00 \qquad\qquad .4 = .40 = .400$$

You can drop zeros at the end of a decimal without changing its value.

$$.500 = .50 = .5 \qquad\qquad 1.000 = 1$$

You must *not* drop zeros in the middle of a decimal. This will change the value of the number. The symbol ≠ means *is not equal to*.

$$\$1.05 \neq \$1.50 \qquad\qquad 6.906 \neq 6.96$$

Use These Steps

Decide if these decimals are equal: 1.30 ☐ 1.3

1. If there is a zero at the end of the decimal, you can drop it.

    1.3̶0̶ ☐ 1.3

2. The decimals are equal. Write = in the box.

    1.3 = 1.3

Decide if the decimals are equal. Write = or ≠ in each box.

1. $6 = $6.00

2. $1.10 ≠ $1.01

3. $4.49 ☐ $40.49

4. 5.6 ☐ 5.600

5. 17.03 ☐ 17.3

6. .8 ☐ .80

7. 100.02 ☐ 1.2

8. 209.0 ☐ 209

9. 9.9 ☐ 90.90

10. 12.300 ☐ 12.3

11. .05 ☐ .50

12. .67 ☐ .670

13. $3.90 ☐ $3.09

14. $ .79 ☐ $7.90

15. $ .99 ☐ $9.09

16. 10 ☐ 10.00

17. 30.0 ☐ 3

18. 100 ☐ 100.00

19. 8.600 ☐ 8.060

20. 35.09 ☐ 35.90

21. .550 ☐ .55

Changing Fractions to Decimals

You can change most fractions to decimals by dividing. In fact, the fractions bar shows division: the numerator is divided by the denominator.

$$\frac{1}{5} = 5\overline{)1} \qquad\qquad \frac{2}{15} = 15\overline{)2} \qquad\qquad \frac{10}{100} = 100\overline{)10}$$

Use These Steps

Change $\frac{1}{4}$ to a decimal.

1. Set up a division problem.

2. Add a decimal point and a zero. Put a decimal point in the answer above the decimal point after the 1.

3. Divide until there is no remainder. Add more zeros if needed.

$$\frac{1}{4} = 4\overline{)1} \qquad\qquad 4\overline{)1.0} \qquad\qquad \begin{array}{r} .25 \\ 4\overline{)1.00} \\ -8 \\ \hline 20 \\ -20 \\ \hline 0 \end{array}$$

Change each fraction to a decimal using division.

1.
$$\frac{1}{5} = .2$$
$$\begin{array}{r} .2 \\ 5\overline{)1.0} \\ -10 \\ \hline 0 \end{array}$$

2. $\frac{67}{100} =$

3. $\frac{9}{10} =$

4. $\frac{3}{4} =$

5. $\frac{3}{20} =$

6. $\frac{17}{25} =$

7. $\frac{11}{20} =$

8. $\frac{4}{5} =$

9. $\frac{1}{2} =$

10. $\frac{2}{5} =$

11. $\frac{22}{25} =$

12. $\frac{4}{25} =$

13. $\frac{3}{5} =$

14. $\frac{7}{20} =$

15. $\frac{13}{25} =$

16. $\frac{37}{100} =$

Changing Fractions to Decimals

When you divide the numerator by the denominator to change some fractions to decimals, your answer may have a remainder. When this happens, write the answer with two digits to the right of the decimal point. Write the remainder as a fraction. Reduce if necessary.

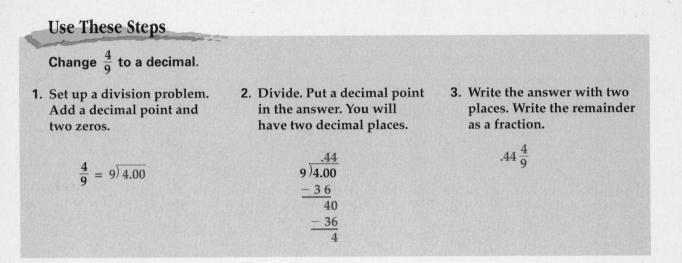

Use These Steps

Change $\frac{4}{9}$ to a decimal.

1. Set up a division problem. Add a decimal point and two zeros.

$$\frac{4}{9} = 9\overline{)4.00}$$

2. Divide. Put a decimal point in the answer. You will have two decimal places.

$$\begin{array}{r} .44 \\ 9\overline{)4.00} \\ -36 \\ \hline 40 \\ -36 \\ \hline 4 \end{array}$$

3. Write the answer with two places. Write the remainder as a fraction.

$.44\frac{4}{9}$

Change each fraction to a decimal. Write the remainder as a fraction. Reduce if possible.

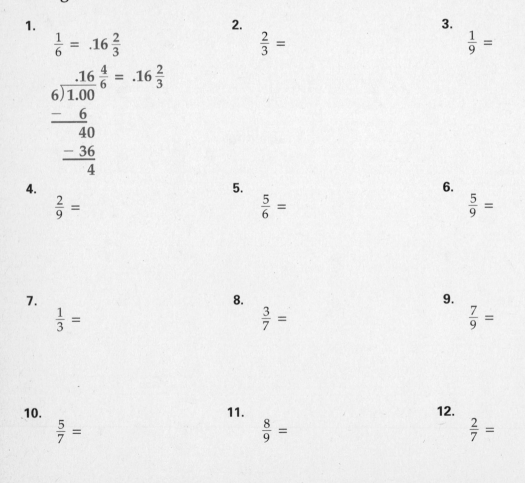

1.
$$\frac{1}{6} = .16\frac{2}{3}$$

$$\begin{array}{r} .16\frac{4}{6} = .16\frac{2}{3} \\ 6\overline{)1.00} \\ -6 \\ \hline 40 \\ -36 \\ \hline 4 \end{array}$$

2.
$$\frac{2}{3} =$$

3.
$$\frac{1}{9} =$$

4.
$$\frac{2}{9} =$$

5.
$$\frac{5}{6} =$$

6.
$$\frac{5}{9} =$$

7.
$$\frac{1}{3} =$$

8.
$$\frac{3}{7} =$$

9.
$$\frac{7}{9} =$$

10.
$$\frac{5}{7} =$$

11.
$$\frac{8}{9} =$$

12.
$$\frac{2}{7} =$$

Comparing Decimals

To compare decimals, line up the decimal points. Start at the left, and compare the value of the digits in each place. The greater number is the number with the greater digit farthest to the left.

Use the symbol < for *less than*.
Use the symbol > for *greater than*.

Comparing decimals is easier if both decimals have the same number of places. Add zeros to the end of the decimal if needed.

.623 > .543   1.03 < 1.037

.6 2 3   1.0 3 0
.5 4 3   1.0 3 7

Use These Steps

Compare .509 ☐ .574

1. Line up the numbers by the decimal points.

2. Compare the digits in the tenths place. 5 = 5. They are the same.

3. Compare the digits in the hundredths place. 0 is less than 7. There is no need to compare the digits in the thousandths place.

.509   .5 09   .5 0 9
.574   .5 74   .5 7 4

0 < 7, so .509 < .574

Compare the following decimals. Write >, <, or = in each box.

1.
.73 < .83
.7 3
.8 3

2.
.9 ☐ .91

3.
.40 ☐ .4

4.
1.362 ☐ .363

5.
25.06 ☐ 25.6

6.
9.09 ☐ .099

7.
3.42 ☐ 3.4

8.
6.92 ☐ 6.92

9.
10.1 ☐ 10.0

10.
.88 ☐ .880

11.
1.033 ☐ 1.33

12.
3.30 ☐ 3.3

13.
.01 ☐ .10

14.
.10 ☐ .100

15.
.11 ☐ 1.1

Real-Life Application

Scales are used to weigh meat and produce. Electronic scales measure weight in hundredths of a pound. These scales print labels using decimals to show the net weight (Net Wt.), price per pound (Price/lb.), and the total price. The net weight is the actual weight of the food, and does not include the weight of the container.

Example Mary asked for $4\frac{1}{2}$ pounds of stew meat. What will the net weight read on the label?

Change the mixed number $4\frac{1}{2}$ to a decimal.

$$4\frac{1}{2} = 4\frac{1 \times 50}{2 \times 50} = 4\frac{50}{100} = 4.50$$

The label will read 4.50.

Solve.

1. Mary ordered $2\frac{1}{4}$ pounds of ground beef. What will the net weight show on the label?

   Answer_____

2. Mary's husband picked out a package of macaroni. The net weight was .9 pound. What fraction of a pound is this?

   Answer_____

3. Mary asked her son Joey to get $\frac{3}{4}$ pound of grapes. What should the label read for $\frac{3}{4}$ pound?

   Answer_____

4. The package label on the grapes Joey bought showed net weight .50. Did Joey buy more or less than $\frac{3}{4}$ pound?

   Answer_____

5. Mary needs about $1\frac{3}{4}$ pounds of cheese to make some pizza. What should the label read for $1\frac{3}{4}$ pounds?

   Answer_____

6. The label on a package of cheese Mary bought showed net weight 1.65. Did Mary buy more or less than $1\frac{3}{4}$ pounds?

   Answer_____

Rounding to the Nearest Whole Number

To round a decimal to the nearest whole number, look at the digit in the tenths place. If it is less than 5, drop the decimal point and all the digits to the right.

> 2.3 rounds to 2
> .45 rounds to 0
> 17.033 rounds to 17

If the digit in the tenths place is 5 or greater, add 1 to the number in the ones place. Drop the decimal point and all the digits to the right.

> .5 rounds to 1
> 4.75 rounds to 5
> 17.622 rounds to 18

Use These Steps

Round 29.63 to the nearest whole number.

1. **Look at the digit in the tenths place.**

   29.<u>6</u>3

2. **Since the digit 6 is greater than 5, add 1 to the whole number, 29. Drop the decimal point and the digits to the right.**

   29.63 rounds to 30

Round each decimal to the nearest whole number.

1. 1.3 ___1___

2. 6.5 _____

3. 2.9 _____

4. 32.54 _____

5. 17.06 _____

6. 93.16 _____

7. 10.632 _____

8. 40.801 _____

9. 50.076 _____

10. .001 _____

11. .93 _____

12. .5 _____

13. 9.9 _____

14. 39.46 _____

15. 99.99 _____

16. 199.39 _____

17. 209.09 _____

18. 304.702 _____

19. 16.505 _____

20. .113 _____

21. 167.50 _____

Rounding to the Nearest Tenth

To round a decimal to the nearest tenth, look at the digit in the hundredths place. If it is less than 5, drop all the digits to the right of the tenths place.

.33 rounds to .3
4.446 rounds to 4.4
17.003 rounds to 17.0

If the digit in the hundredths place is 5 or greater, add 1 to the number in the tenths place. Drop all digits to the right of the tenths place.

2.56 rounds to 2.6
4.759 rounds to 4.8
.082 rounds to .1

Use These Steps

Round .693 to the nearest tenth.

1. Look at the digit in the hundredths place.

.6_9_3

2. Since the digit 9 is greater than 5, add 1 to the digit in tenths place, 6. Drop the digits to the right.

.693 rounds to .7

Round each decimal to the nearest tenth.

1. 1.73   1.7

2. 7.86   _____

3. 2.91   _____

4. 2.55   _____

5. 7.08   _____

6. 3.16   _____

7. 10.63   _____

8. 40.80   _____

9. 50.07   _____

10. .051   _____

11. .906   _____

12. .591   _____

13. 9.970   _____

14. 39.946   _____

15. 99.999   _____

16. 199.739   _____

17. 209.099   _____

18. 999.909   _____

19. .04   _____

20. .011   _____

21. 1.04   _____

Rounding to the Nearest Hundredth

To round a decimal to the nearest hundredth, look at the digit in the thousandths place. If it is less than 5, drop all the digits to the right of the hundredths place.

$$2.331 \text{ rounds to } 2.33$$
$$.4846 \text{ rounds to } .48$$
$$17.0033 \text{ rounds to } 17.00$$

If the digit in the thousandths place is 5 or greater, add 1 to the number in the hundredths place. Drop all digits to the right of the hundredths place.

$$.565 \text{ rounds to } .57$$
$$4.7599 \text{ rounds to } 4.76$$
$$1.0072 \text{ rounds to } 1.01$$

Use These Steps

Round .134 to the nearest hundredth.

1. Look at the digit in the thousandths place.

.134

2. Since the digit 4 is less than 5, drop the digits to the right of the hundredths place.

.134 rounds to .13

Round each decimal to the nearest hundredth.

1. 7.348   _7.35_

2. 6.835   _____

3. 2.901   _____

4. 32.005   _____

5. 17.049   _____

6. 93.167   _____

7. 10.632   _____

8. 40.801   _____

9. 50.076   _____

10. .0301   _____

11. .9330   _____

12. .5555   _____

13. 8.099   _____

14. 39.496   _____

15. 99.999   _____

16. 199.3909   _____

17. 209.0987   _____

18. .9999   _____

19. 2.0041   _____

20. 2.0046   _____

21. 2.650   _____

Real-Life Application   At Work

Abby is a quality-control worker. She works in the produce packaging department of Bert's Super Store. Her job is to inspect packages. She checks the weight of every tenth package. She compares the weight of each package to a standard weight.

Abby fills in a chart with the weight of each package. If the package weighs too much, she puts a + in her chart. If it weighs too little, she writes a − on her chart. If it weighs the same as the standard she writes =.

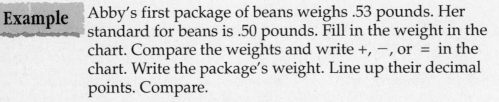

Example Abby's first package of beans weighs .53 pounds. Her standard for beans is .50 pounds. Fill in the weight in the chart. Compare the weights and write +, −, or = in the chart. Write the package's weight. Line up their decimal points. Compare.

$$. 5 \boxed{3}$$
$$. 5 \boxed{0}$$
$$3 > 0, \text{ so } .53 > .50.$$

Since the package weight, .53, is greater than the standard, .50, write + in the chart.

	Example	1	2	3	4	5	6	7	8	9	10	11	12
Weight	.53	.49											
+, −, =	+	−											

Complete the chart using the package weights below.

1. .49   2. .66   3. .46   4. .39

5. .51   6. .65   7. .50   8. .44

9. .52   10. .47   11. .59   12. .64

Unit 1 *Review*

Change the fractions to decimals.

1.
$\frac{1}{2}$

2.
$\frac{3}{4}$

3.
$\frac{3}{5}$

4.
$\frac{17}{25}$

5.
$\frac{9}{10}$

6.
$\frac{13}{50}$

Write a fraction or mixed number. Then change to a decimal.

7.
two tenths

8.
two hundredths

9.
ten and four tenths

10.
four and no hundredths

11.
twenty and six hundredths

Change each decimal to a fraction or mixed number. Reduce if possible.

12.
.08

13.
.3

14.
.14

15.
.27

16.
9.79

17.
10.66

18.
74.2

19.
6.5

Write these amounts as decimals.

20.
ninety-five cents

21.
one dollar and six cents

22.
seven dollars

Write each amount in words.

23. $ .60 _____

24. $207.25 _____

25. $28.00 _____

26. $13.03 _____

Put each number in the chart. Then write each number in words.

27. 1.09 _____

28. 5.3 _____

29. 350.05 _____

30. .45 _____

31. .225 _____

	hundreds	tens	ones	.	tenths	hundredths	thousandths
27.				.			
28.				.			
29.				.			
30.				.			
31.				.			

Decide if the decimals are equal. Write = or ≠ in each box.

32.
  $.81 ☐ $8.10

33.
  400 ☐ 400.00

34.
  6.90 ☐ 6.09

35.
  $53.00 ☐ $5.30

Change each fraction to a decimal. Write the remainder as a fraction. Reduce if possible.

36.
  $\frac{4}{7}$ =

37.
  $\frac{7}{9}$ =

38.
  $\frac{1}{6}$ =

Compare each set of numbers. Write =, <, or > in each box.

39.
  .8 ☐ .80

40.
  .75 ☐ .76

41.
  .40 ☐ .44

42.
  33.09 ☐ 3.309

Round to the nearest whole number.

43.
  10.09

44.
  7.50

45.
  .59

46.
  312.2

47.
  784.29

Round to the nearest tenth.

48.
  .95

49.
  19.36

50.
  109.99

51.
  302.11

52.
  4.14

Round to the nearest hundredth.

53.
  1.001

54.
  .0999

55.
  3.155

56.
  989.002

57.
  20.101

Below is a list of the problems in this review and the pages on which the skills are taught. If you missed any problems, turn to the pages listed and practice the skills. Then correct the problems you missed in the Unit Review.

You add and subtract decimals whenever you add and subtract money. Some examples of using decimals and money are adding up purchases, counting your change from a purchase, and balancing a checkbook.

Adding and subtracting decimals is the same as adding and subtracting whole numbers. You must be sure, however, to line up the decimal points so that you are always adding or subtracting digits with the same place value.

In this unit, you will learn how to add and subtract money and other decimals.

Getting Ready

You should be familiar with the skills on this page and the next before you begin this unit. To check your answers, turn to page 165.

▶ To add or subtract whole numbers, line up the digits that have the same place value.

Add or subtract.

1.
$$462 + 22 =$$
$$\begin{array}{r} 462 \\ +\ 22 \\ \hline 484 \end{array}$$

2.
$$948 - 217 =$$

3.
$$67,499 - 3,372 =$$

4.
$$8,142 + 656 =$$

5.
$$15,948 - 932 =$$

6.
$$94,162 + 3,115 =$$

For review, see pages 34–37, 60–62 in **Math Matters for Adults, Whole Numbers.**

Sometimes when adding two or more digits, the total is ten or more. When this happens, rename by carrying tens to the next column or the next place to the left.

Add.

7.
$$95 + 37 =$$

$$\begin{array}{r} 1 \\ 95 \\ +\ 37 \\ \hline 132 \end{array}$$

8.
$$1,147 + 283 =$$

9.
$$942 + 76 + 153 =$$

10.
$$\begin{array}{r} 39 \\ +\ 41 \end{array}$$

11.
$$\begin{array}{r} 479 \\ +\ 67 \end{array}$$

12.
$$\begin{array}{r} 2,431 \\ +\ 988 \end{array}$$

13.
$$\begin{array}{r} 15,889 \\ +\ 1,473 \end{array}$$

14.
$$\begin{array}{r} 62,437 \\ 97,563 \\ +\ 3,259 \end{array}$$

For review, see pages 41–44 in **Math Matters for Adults, Whole Numbers.**

Sometimes when subtracting, the digit you are subtracting is larger than the digit you are subtracting from. When this happens, rename by borrowing from the next column to the left.

Subtract.

15.
$$715 - 49 =$$

$$\begin{array}{r} 10 \\ 6\ \cancel{0}15 \\ 7\cancel{1}\cancel{5} \\ -\ 49 \\ \hline 666 \end{array}$$

16.
$$436 - 55 =$$

17.
$$1,462 - 999 =$$

18.
$$\begin{array}{r} 62 \\ -\ 59 \end{array}$$

19.
$$\begin{array}{r} 837 \\ -148 \end{array}$$

20.
$$\begin{array}{r} 1,990 \\ -\ 876 \end{array}$$

21.
$$\begin{array}{r} 42,493 \\ -\ 1,294 \end{array}$$

22.
$$\begin{array}{r} 137,491 \\ -\ 42,586 \end{array}$$

For review, see pages 72–73 in **Math Matters for Adults, Whole Numbers.**

Sometimes you have to add or subtract zeros. Zero plus any number is always that number. When subtracting from a number with zero, you may need to rename.

Add or subtract.

23.
$$\begin{array}{r} 250 \\ +\ 300 \\ \hline 550 \end{array}$$

24.
$$\begin{array}{r} 4,700 \\ +\ 2,000 \end{array}$$

25.
$$\begin{array}{r} 50 \\ -\ 25 \end{array}$$

26.
$$\begin{array}{r} 701 \\ -\ 32 \end{array}$$

27.
$$\begin{array}{r} 1,000 \\ -\ 978 \end{array}$$

For review, see pages 45, 68–69, 75–78 in **Math Matters for Adults, Whole Numbers.**

Adding Money

When you add amounts of money, be sure the decimal points are lined up. Line up the decimal point in the answer with those in the problem.

$$\$2.39 + \$8.40 \qquad \begin{array}{r} \$\ 2.39 \\ +\ \ \ 8.40 \\ \hline \$10.79 \end{array}$$

Use These Steps

Add $6.33 + $4.95

1. Be sure the decimal points are lined up.

$$\begin{array}{r} \$6.33 \\ +\ \ 4.95 \\ \hline \end{array}$$

2. Add. Begin with the digits on the right. Rename. Put a decimal point and a dollar sign in the answer.

$$\begin{array}{r} 1\ \ \ \ \\ \$\ 6.33 \\ +\ \ 4.95 \\ \hline \$11.28 \end{array}$$

Add. Rename if necessary.

1.
$$\$6.75 + \$2.31 =$$
$$\begin{array}{r} 1\ \ \ \ \\ \$6.75 \\ +\ \ 2.31 \\ \hline \$9.06 \end{array}$$

2. $\$\ .40 + \$\ .66 =$

3. $\$\ .99 + \$1.84 =$

4. $\$\ .40 + \$\ .95 + \$1.30 =$

5. $\$3.99 + \$2.97 + \$4.67 =$

6. $\$6.69 + \$4.03 =$

7. $\$3.06 + \$5.60 =$

8. $\$7.99 + \$1.99 =$

9. $\$8.42 + \$6.11 + \$3.99 =$

10. $\$1.49 + \$\ .99 + \$4.08 =$

11. $\$\ .06 + \$1.29 + \$5.88 + \$\ .39 =$

12. $\$2.59 + \$\ .95 + \$\ .33 + \$4.41 =$

Adding Money

When you add whole dollar amounts of money, include a decimal point and two zeros in the tenths and hundredths places.

$39.41 + $6

$$
\begin{array}{r}
\overset{1}{} \\
\$39.41 \\
+\quad 6.00 \\
\hline
\$45.41
\end{array}
$$

⟵ add two zeros

Use These Steps

Add $9.09 + $17

1. Set up the problem by lining up the decimal points, one above the other. Add a decimal point and zeros.

$$
\begin{array}{r}
\$\ 9.09 \\
+\ 17.00 \\
\hline
\end{array}
$$

⟵ add two zeros

2. Add. Begin with the digits on the right. Rename. Put a decimal point and a dollar sign in the answer.

$$
\begin{array}{r}
\overset{1}{} \\
\$\ 9.09 \\
+\ 17.00 \\
\hline
\$26.09
\end{array}
$$

Add. Rename if necessary.

1.

$650 + $73.29 =

$$
\begin{array}{r}
\overset{1}{} \\
\$650.00 \\
+\quad 73.29 \\
\hline
\$723.29
\end{array}
$$

2.

$44.40 + $7 =

3.

$760.02 + $99 =

4.

$5.99 + $19 + $400 =

$$
\begin{array}{r}
\overset{1}{} \\
\$\quad 5.99 \\
19.00 \\
+\ 400.00 \\
\hline
\$424.99
\end{array}
$$

5.

$340 + $66.09 + $ .78 =

6.

$342.93 + $61.40 + $100 =

7.

$9 + $ .75 + $499 =

8. Lee bought a shirt for $10 and a pair of pants for $29.99. What was the total cost of the clothes he bought?

9. Lee bought a pair of shoes for $35. The sales tax was $2.80. Including the tax, how much did he spend on the shoes?

Answer _____

Answer _____

Adding Decimals

A whole number can be written as a decimal by placing a decimal point followed by one or more zeros to the right of the decimal point.

To add decimals, line up the decimal points first. Add zeros and rename if necessary. Be sure to line up the decimal point in your answer with the other decimal points in the problem.

$8 + 2.66$

$$
\begin{array}{r}
8.00 \quad \longleftarrow \text{add a decimal point and two zeros} \\
+ 2.66 \\
\hline
10.66
\end{array}
$$

Use These Steps

Add 1.8 + 4.25

1. Set up the problem by lining up the decimal points. Add a zero.

$$
\begin{array}{r}
1.80 \quad \longleftarrow \text{add a zero} \\
+ 4.25 \\
\hline
\end{array}
$$

2. Add. Begin with the digit on the right. Rename. Put a decimal point in the answer.

$$
\begin{array}{r}
\overset{1}{1}.80 \\
+ 4.25 \\
\hline
6.05
\end{array}
$$

Add. Rename if necessary.

1.
$3 + 42.6 =$

$$
\begin{array}{r}
3.0 \\
+ 42.6 \\
\hline
45.6
\end{array}
$$

2.
$99.9 + 49 =$

3.
$77.3 + 1.97 =$

4.
$1 + 44.6 =$

5.
$82 + 199 + 11.03 =$

6.
$.001 + .01 + 1.1 =$

7.
$5 + 105 + 5.01 =$

8.
$$
\begin{array}{r}
6.3 \\
+ 7.9 \\
\hline
\end{array}
$$

9.
$$
\begin{array}{r}
13.04 \\
+ 6.96 \\
\hline
\end{array}
$$

10.
$$
\begin{array}{r}
1.001 \\
+ 9.009 \\
\hline
\end{array}
$$

11.
$$
\begin{array}{r}
19.3 \\
+ 24.7 \\
\hline
\end{array}
$$

12.
$$
\begin{array}{r}
.72 \\
+ .57 \\
\hline
\end{array}
$$

13.
$$
\begin{array}{r}
100.3 \\
+ 42.75 \\
\hline
\end{array}
$$

14.
$$
\begin{array}{r}
89.73 \\
+ 4.001 \\
\hline
\end{array}
$$

15.
$$
\begin{array}{r}
498.032 \\
6.047 \\
+ 10.93 \\
\hline
\end{array}
$$

16.
$$
\begin{array}{r}
7.4 \\
7.9 \\
+ 10.03 \\
\hline
\end{array}
$$

17.
$$
\begin{array}{r}
143.99 \\
1.01 \\
+ 10. \\
\hline
\end{array}
$$

Adding Decimals

When adding decimals, be sure to line up all of the decimal points, including the one in your answer. Rename if necessary. Add zeros to make the same number of decimal places.

.5732 + 19.032

$$\begin{array}{r} 1 \\ .5732 \\ + 19.0320 \quad \longleftarrow \text{add a zero} \\ \hline 19.6052 \end{array}$$

Use These Steps

Add 75.0327 + 850.3

1. Set up the problem by lining up the decimal points. Add zeros.

$$\begin{array}{r} 75.0327 \\ + 850.3000 \quad \longleftarrow \text{add three zeros} \\ \hline \end{array}$$

2. Add. Begin with the digits on the right. Rename when needed. Put a decimal point in the answer.

$$\begin{array}{r} 1 \\ 75.0327 \\ + 850.3000 \\ \hline 925.3327 \end{array}$$

Add. Rename if necessary.

1.
$$\begin{array}{r} 1 \\ .730 \\ + .079 \\ \hline .809 \end{array}$$

2.
$$\begin{array}{r} .604 \\ + 1.3069 \\ \hline \end{array}$$

3.
$$\begin{array}{r} 11. \\ + 9.309 \\ \hline \end{array}$$

4.
$$\begin{array}{r} 81.1055 \\ + .34 \\ \hline \end{array}$$

5.
$$\begin{array}{r} 117.7 \\ + 96.037 \\ \hline \end{array}$$

6.
$$\begin{array}{r} 14.5933 \\ .6 \\ + 1.01 \\ \hline \end{array}$$

7.
$$\begin{array}{r} 49.3211 \\ 7.19 \\ + 3. \\ \hline \end{array}$$

8.
$$\begin{array}{r} .4499 \\ 3.3 \\ + 12.0729 \\ \hline \end{array}$$

9.
$$\begin{array}{r} 5.0555 \\ 137.05 \\ + 1.1 \\ \hline \end{array}$$

10.
$$\begin{array}{r} 10. \\ .6 \\ + 3.999 \\ \hline \end{array}$$

11.
4 + 10.6 + .325 =
$$\begin{array}{r} 4.000 \\ 10.600 \\ + .325 \\ \hline 14.925 \end{array}$$

12.
785.09 + 1 + .5 =

13.
12.01 + 6 + 3.7 =

14.
19.0445 + 23 + .889 =

15.
34.6789 + 545.338 + 2.5678 =

Real-Life Application

Macy is buying school supplies for her children. She is rounding to the nearest dollar the price of each item she buys so that she can stay within her budget.

Example   Macy wants to buy a box of pencils for $1.19 and a pack of paper for $1.69. About how much money will she need?

$1.19 rounds to $1.00 $1.00
$1.69 rounds to $2.00   + 2.00
                             $3.00

Macy will need about $3.00.

Solve.

1. Macy wants to buy a pen for $2.78 and a notebook for $ .89. About how much money will she need?

Answer_____

2. Macy stopped at the lunch counter. She bought a sandwich for $1.79 and a drink for $ .75. About how much did she spend for lunch?

Answer_____

3. Macy wants to buy a backpack for $10.25 and a watch for $19.95. About how much money will she need?

Answer_____

4. Macy wants to buy a paperback dictionary for $6.95 and a calculator for $7.59. About how much money will she need?

Answer_____

Mixed Review

Change each fraction to a decimal.

1. $\dfrac{5}{10} =$

2. $\dfrac{16}{100} =$

3. $\dfrac{5}{20} =$

4. $\dfrac{7}{25} =$

5. twelve hundredths =

6. four fifths =

Add. Rename if necessary.

7. $\$7.50 + \$3.50 =$

8. $\$56.20 + \$21.10 =$

9. $\$109.26 + \$96.04 =$

10. $\$ .10 + \$3.01 + \$6.17 =$

11. $\$11.26 + \$ .98 + \$102.06 =$

12. $\$4 + \$6.90 =$

13. $\$6 + \$15.89 =$

14. $\$110 + \$81.29 =$

15.
$$
\begin{array}{r}
5.09 \\
+\ 6.3 \\
\hline
\end{array}
$$

16.
$$
\begin{array}{r}
7.12 \\
+\ 9.34 \\
\hline
\end{array}
$$

17.
$$
\begin{array}{r}
18.04 \\
+\ 25.7 \\
\hline
\end{array}
$$

18.
$$
\begin{array}{r}
.13 \\
+\ 2.09 \\
\hline
\end{array}
$$

19.
$$
\begin{array}{r}
2.0113 \\
+\ 7.5 \\
\hline
\end{array}
$$

20.
$$
\begin{array}{r}
.093 \\
.6 \\
+\ 25.07 \\
\hline
\end{array}
$$

21.
$$
\begin{array}{r}
50.139 \\
1.82 \\
+\ \ \ .011 \\
\hline
\end{array}
$$

22.
$$
\begin{array}{r}
.301 \\
115.12 \\
+\ 320.909 \\
\hline
\end{array}
$$

23.
$$
\begin{array}{r}
273.01 \\
134.131 \\
+\ \ \ 6.81 \\
\hline
\end{array}
$$

24.
$$
\begin{array}{r}
497.263 \\
50.001 \\
+\ 148.294 \\
\hline
\end{array}
$$

25. $6.3 + 4.7 =$

26. $8 + 76.2 + .18 =$

27. $58 + 101.29 + 2.7 =$

Subtracting Money

To subtract money, first line up the decimal points. Then subtract.
Rename if necessary. Remember to put a dollar sign in the answer and
to line up the decimal point in the answer with those in the problem.

$$\$5.96 - \$3.48 \qquad \begin{array}{r} {\scriptstyle 8\ 16} \\ \$5.9\llap{/}6 \\ -\ \ 3.48 \\ \hline \$2.48 \end{array}$$

Use These Steps

Subtract $9.40 − $5.49

1. Set up the problem by lining up the decimal points.

2. Subtract. Begin with the digits on the right. Rename. Put a decimal point and a dollar sign in the answer.

$$\begin{array}{r} \$9.40 \\ -\ 5.49 \\ \hline \end{array} \qquad \begin{array}{r} {\scriptstyle 13} \\ {\scriptstyle 8\ 3\ 10} \\ \$9.4\,0 \\ -\ 5.4\,9 \\ \hline \$3.9\,1 \end{array}$$

Subtract. Rename if necessary.

1.
$$\$4.29 - \$2.18 =$$
$$\begin{array}{r} \$4.29 \\ -\ \ 2.18 \\ \hline \$2.11 \end{array}$$

2.
$$\$8.88 - \$\ .52 \ =$$

3.
$$\$5.88 - \$4.12 =$$

4.
$$\$9.01 - \$3.14 =$$
$$\begin{array}{r} {\scriptstyle 9} \\ {\scriptstyle 8\ 10\ 11} \\ \$9.0\,1 \\ -\ \ 3.1\,4 \\ \hline \$5.8\,7 \end{array}$$

5.
$$\$\ .62 - \$\ .26 =$$

6.
$$\$8.46 - \$\ .98 =$$

7.
$$\$5.39 - \$2.03 =$$

8.
$$\$8.77 - \$\ .93 =$$

9.
$$\$9.06 - \$7.96 =$$

10.
$$\$7.20 - \$2.75 =$$

11.
$$\$10.02 - \$4.99 =$$

12.
$$\$1.03 - \$\ .56 =$$

Subtracting Money

When subtracting from a whole dollar amount, be sure to add a decimal point and two zeros. Rename if necessary.

$$\$15 - \$3.97 \qquad \begin{array}{r} {\scriptstyle 9} \\ {\scriptstyle 4\,10\,10} \\ \$1\,5.0\,0 \\ -\quad 3.9\,7 \\ \hline \$1\,1.0\,3 \end{array} \longleftarrow \text{add a decimal point and two zeros}$$

Use These Steps

Subtract $13 − $.67

1. Set up the problem by lining up the decimal points. Add a decimal point and zeros.

$$\begin{array}{r} \$13.00 \\ -\quad .67 \\ \hline \end{array} \longleftarrow \text{add a decimal point and two zeros}$$

2. Subtract. Begin with the digits on the right. Rename. Put a decimal point and a dollar sign in the answer.

$$\begin{array}{r} {\scriptstyle 9} \\ {\scriptstyle 2\,10\,10} \\ \$1\,3.0\,0 \\ -\quad .6\,7 \\ \hline \$1\,2.3\,3 \end{array}$$

Subtract. Rename if necessary.

1.
$$\$20 - \$5.67 = \qquad \begin{array}{r} {\scriptstyle 9\;\; 9} \\ {\scriptstyle 1\,10\,10\,10} \\ \$2\,0.0\,0 \\ -\quad 5.6\,7 \\ \hline \$1\,4.3\,3 \end{array}$$

2.
$$\$38 - \$6.13 =$$

3.
$$\$120 - \$.99 =$$

4.
$$\$106 - \$39.55 =$$

5.
$$\$382 - \$134.01 =$$

6.
$$\$486 - \$277.66 =$$

7.
$$\$300 - \$149.95 =$$

8.
$$\$1,000 - \$962.58 =$$

9.
$$\$4,000 - \$247.50 =$$

10. Sean had $20. He bought a magazine for $1.50. How much money does he have left?

11. Jules paid $10.75 for a money order. He gave the cashier $15. How much change did Jules get?

Answer _____

Answer _____

Subtracting Decimals

To subtract decimals, first set up the problem by lining up the decimal points just as with money.

Add a decimal point and zeros if needed so that both decimals have the same number of places after the decimal point. Then subtract. Line up the decimal point in the answer with the other decimal points. Rename if necessary.

$$
10 - 9.3 \qquad
\begin{array}{r}
\overset{9}{\underset{}{\cancel{1}\,\cancel{0}.\cancel{0}}} \\[-2pt]
\scriptstyle 0\,\cancel{10}\,10 \\
\end{array}
$$

$$
\begin{array}{r}
1\,0.0 \\
-\ \ 9.3 \\
\hline
.7
\end{array}
$$ ◄── add a decimal point and a zero

Use These Steps

Subtract 14.06 − 1.325

1. Set up the problem by lining up the decimal points. Add a zero.

$$
\begin{array}{r}
14.060 \\
-\ \ 1.325 \\
\hline
\end{array}
$$ ◄── add a zero

2. Subtract. Begin with the digits on the right. Rename. Put a decimal point in the answer.

$$
\begin{array}{r}
\scriptstyle 3\ 10\ 5\ 10 \\
1\,4.0\,6\,0 \\
-\ \ 1.3\,2\,5 \\
\hline
1\,2.7\,3\,5
\end{array}
$$

Subtract. Rename if necessary.

1.
$$
\begin{array}{r}
\scriptstyle 8\ 10 \\
3.9\cancel{0} \\
-\ 2.63 \\
\hline
1.27
\end{array}
$$

2.
$$
\begin{array}{r}
8.3 \\
-\ 5.21 \\
\hline
\end{array}
$$

3.
$$
\begin{array}{r}
2. \\
-\ 1.01 \\
\hline
\end{array}
$$

4.
$$
\begin{array}{r}
.1 \\
-\ .09 \\
\hline
\end{array}
$$

5.
$$
\begin{array}{r}
1. \\
-\ .13 \\
\hline
\end{array}
$$

6.
$$
\begin{array}{r}
16.3 \\
-\ 2.81 \\
\hline
\end{array}
$$

7.
$$
\begin{array}{r}
9. \\
-\ .3 \\
\hline
\end{array}
$$

8.
$$
\begin{array}{r}
4. \\
-\ 3.42 \\
\hline
\end{array}
$$

9.
$$
\begin{array}{r}
9.04 \\
-\ 1.035 \\
\hline
\end{array}
$$

10.
$$
\begin{array}{r}
.72 \\
-\ .349 \\
\hline
\end{array}
$$

11. $8.26 - 4.6 =$

12. $3.002 - .01 =$

13. $8 - 5.09 =$

14. $1 - .873 =$

15. $9.3 - 3.44 =$

16. $1 - .044 =$

17. $2.01 - 1.024 =$

18. $4 - .241 =$

19. $.6 - .455 =$

20. $.78 - .194 =$

Subtracting Decimals

When subtracting decimals, be sure to line up the decimal points. Add a decimal point and zeros when needed. Rename if necessary.

$$1003 - 956.473$$

$$
\begin{array}{r}
9\ \ 9\ 12\ 9\ \ 9 \\
0\ 10\ 10\ \ 2\ 10\ 10\ 10 \\
1\,0\,0\,3.0\,0\,0 \\
-\ \ 9\,5\,6.4\,7\,3 \\
\hline
4\,6.5\,2\,7
\end{array}
$$

← add a decimal point and three zeroes

Use These Steps

Subtract 275.5 − 198.75

1. Set up the problem by lining up the decimal points. Add a zero.

$$
\begin{array}{r}
275.50 \\
-\ 198.75 \\
\hline
\end{array}
$$

← add a zero

2. Subtract. Begin with the digits on the right. Rename. Put a decimal point in the answer.

$$
\begin{array}{r}
16\ 14\ 14 \\
1\ \ 6\ \ 4\ \ 4\ 10 \\
2\ 7\ 5.5\ 0 \\
-1\ 9\ 8.7\ 5 \\
\hline
7\ 6.7\ 5
\end{array}
$$

Subtract. Rename if necessary.

1.
$$
\begin{array}{r}
12 \\
7\ \ 2\ 10 \\
.8\ 3\ 0 \\
-.1\ 3\ 1 \\
\hline
.6\ 9\ 9
\end{array}
$$

2.
$$
\begin{array}{r}
8.612 \\
-\ 1.44 \\
\hline
\end{array}
$$

3.
$$
\begin{array}{r}
20. \\
-\ 8.17 \\
\hline
\end{array}
$$

4.
$$
\begin{array}{r}
86.21 \\
-\ \ .3 \\
\hline
\end{array}
$$

5.
$$
\begin{array}{r}
115.1 \\
-\ 27.89 \\
\hline
\end{array}
$$

6.
$$
\begin{array}{r}
82. \\
-\ 3.67 \\
\hline
\end{array}
$$

7.
$$
\begin{array}{r}
103.1 \\
-\ 17.12 \\
\hline
\end{array}
$$

8.
$$
\begin{array}{r}
93.28 \\
-\ .391 \\
\hline
\end{array}
$$

9.
$$
\begin{array}{r}
931.001 \\
-\ 50.01 \\
\hline
\end{array}
$$

10.
$$
\begin{array}{r}
810.0 \\
-\ 23.77 \\
\hline
\end{array}
$$

11. $8 - 7.26 =$

12. $36.4 - .59 =$

13. $300 - 1.031 =$

14. $2 - .006 =$

15. Tony needs to work 40 hours this week. So far he has worked 27.5 hours. How many more hours does he need to work?

16. Tony works 8 hours a day. If he has worked 6.29 hours so far today, how many more hours does he have to work?

Answer_____

Answer_____

Problem Solving: Using a Table

The table below shows the income of people based on how many years of school they've completed. For example, of those people who have an elementary education, 4,945 thousand have an income of less than $10,000 a year.

AMOUNT OF EDUCATION (Numbers in thousands)	INCOME LEVEL					
	Under $10,000	$10,000–$14,999	$15,000–$24,999	$25,000–$34,999	$35,000–$49,999	$50,000 and over
Elementary	4,945	1,978	2,265.5	1,138.5	736	425.5
High School	7,879.1	4,924.4	8,946	7,304.6	6,976.3	5,047.6
College (one or more years)	2,297.8	1,931.5	5,028.5	5,428.1	7,559.3	11,122.5

Example   What is the total number of people who have a high school education and receive an income of $25,000 or more a year?

▶ **Step 1.**   Find the row that shows the numbers of people with a high school education.

▶ **Step 2.**   Add the number of people with a high school education in the columns $25,000–$34,999, $35,000–$49,999, and $50,000 and over to find the total number of people receiving an income of $25,000 or more.

$$
\begin{array}{r}
{\scriptstyle 1\ 1\ 1\ 1} \\
7,304.6 \\
6,976.3 \\
+\ 5,074.6 \\
\hline
19,355.5
\end{array}
$$

19,355.5 thousand people with a high school education have an income of $25,000 or more a year.

Solve.

1. What is the total number of people with an elementary education who earn $25,000 or more a year?

2. What is the total number of people with one or more years of college who earn $25,000 or more a year?

Answer_____

Answer_____

3. How many more people with a high school education earn $15,000–$24,999 than earn under $10,000?

Answer_____

4. Which income column lists the greatest number of people with an elementary education?

Answer_____

5. Which income column lists the greatest number of people with a high school education?

Answer_____

6. What is the total number of people who earn less than $10,000 a year?

Answer_____

7. What is the total number of people who earn $50,000 and over a year?

Answer_____

8. What is the total number of people with an elementary education who earn $14,999 or less a year?

Answer_____

9. What is the total number of people with a high school education who earn $14,999 or less a year?

Answer_____

10. What is the total number of people with one or more years of college who earn $14,999 or less a year?

Answer_____

11. Which income column shows the greatest number of people with one or more years of college?

Answer_____

12. Do most people with an income of $25,000–$34,999 have an elementary education, high school education, or one or more years of college?

Answer_____

Unit 2 *Review*

Add. Rename if necessary.

1.
$8.26 + $3.18 =

2.
$30.40 + $24.77 =

3.
$120.34 + $18.25 =

4.
$ .25 + $3.50 + $8.75 =

5.
$103.25 + $98.09 + $201.60 =

6.
$5 + $7.70 =

7.
$8.25 + $11 =

8.
$103.36 + $180 =

9.
```
  6.39
+ 2.81
```

10.
```
  8.36
+ 5.2
```

11.
```
  30.18
+ 26.88
```

12.
```
  81.001
+  3.7
```

13.
```
  102.113
+  37.32
```

14.
```
   .83
   .29
+ 13.629
```

15.
```
  83.316
   2.1
+  9.699
```

16.
```
  263.101
   77.92
+  81.98
```

17.
```
   .312
   31.8
+ 729.299
```

18.
```
  426.813
   59.93
+ 186.399
```

19.
2.3 + .9 + .38 =

20.
10 + 81.7 + 8.7 =

21.
190 + 26.39 + 3.9 =

22.
.1556 + 32 + 1.9 =

23.
57.9921 + 543.02 + 6 =

24.
125.4 + .001 + .0378 =

Subtract. Rename if necessary.

25.
$8.26 − $3.13 =

26.
$13.74 − $4.54 =

27.
$23.11 − $16.12 =

28.
$121.81 − $76.15 =

29.
$183.12 − $28.06 =

30.
$118.59 − $94.63

31.
$6 − $1.89 =

32.
$410 − $339.31 =

33.
$28 − $12.38 =

34.
$$10.001$$
$$- 5.31$$

35.
$$3.8$$
$$- 1.95$$

36.
$$16.02$$
$$- 9.9$$

37.
$$76.01$$
$$- .133$$

38.
$$11.384$$
$$- 9.499$$

39.
$$26.03$$
$$- 18.3$$

40.
$$113.021$$
$$- 1.139$$

41.
$$123.09$$
$$- 99.1$$

42.
$$499.$$
$$- 28.991$$

43.
$$233.1$$
$$- 129.61$$

44.
26.7 − 18.3 =

45.
101 − 76.23 =

46.
111.28 − 23.98 =

47.
50 − .0032 =

48.
6 − .093 =

49.
.073 − .06154 =

Below is a list of the problems in this review and the pages on which the skills are taught. If you missed any problems, turn to the pages listed and practice the skills. Then correct the problems you missed in the Unit Review.

Problems	Pages	Problems	Pages
1-5	33	25-30	39
6-8	34	31-33	40
9-24	35-36	34-49	41-42

You can multiply and divide decimals to figure out your gas mileage, your total costs, or the monthly payments on your charge cards.

Multiplying and dividing decimals is done the same way as multiplying and dividing whole numbers. You must be careful, however, to put decimal points in the correct place in the answers.

Getting Ready

You should be familiar with the skills on this page and the next before you begin this unit. To check your answers, turn to page 170.

> When you are working with decimals, place value is important in setting up problems, renaming, and lining up answers.

Write the value of the underlined digit in each number.

1.
1.2̲4   **2 tenths** _____

2.
.07̲9 _____

3.
42.93̲ _____

4.
5.506̲ _____

5.
346.82̲85 _____

6.
.0̲013 _____

For review, see Unit 1, page 19.

> When setting up problems for multiplication, be sure to line up the digits in the correct columns.

Multiply.

7.
$32 \times 5 =$

$$\begin{array}{r} 1 \\ 32 \\ \times\ \ 5 \\ \hline 160 \end{array}$$

8.
$406 \times 28 =$

9.
$1,029 \times 87 =$

10.
$2,188 \times 346 =$

For review, see pages 89–92, 95–98, 105–113 in **Math Matters for Adults, Whole Numbers.**

> When setting up problems for division, be sure to put the number you are dividing by in front of the $)$.

Divide. Write remainders as fractions in lowest terms.

11.
$44 \div 6 =$

$$\begin{array}{r} 7\frac{2}{6} = 7\frac{1}{3} \\ 6\overline{)44} \\ -42 \\ \hline 2 \end{array}$$

12.
$800 \div 31 =$

13.
$2,403 \div 24 =$

14.
$6,498 \div 722 =$

For review, see pages 130–131, 134–137, 140–142, 147–148 in **Math Matters for Adults, Whole Numbers.**

> When multiplying and dividing decimals, you may need to round your answers.

Round to the nearest whole number.

15.
$1.45 = 1$

16.
$.521 =$

17.
$17.9 =$

18.
$10.04 =$

19.
$.06 =$

Round to the nearest tenth.

20.
$2.09 = 2.1$

21.
$5.77 =$

22.
$20.241 =$

23.
$.953 =$

24.
$45.81 =$

Round to the nearest hundredth.

25.
$5.033 = 5.03$

26.
$.489 =$

27.
$1.001 =$

28.
$14.615 =$

29.
$200.0004 =$

For review, see Unit 1, pages 25–27.

Multiplying Money

When you multiply money by a whole number, first set up the problem by lining up the digits starting at the right. Place the decimal point in your answer so that there are two places to the right of the decimal point. Include a dollar sign in the answer.

$$\$6.25 \times 3 = \quad \begin{array}{r} 1 \\ \$6.25 \\ \times \quad 3 \\ \hline \$18.75 \end{array} \longleftarrow \text{two decimal places}$$

18 dollars and 75 cents

Use These Steps

Multiply $4.31 × 4

1. Set up the problem.

$$\begin{array}{r} \$4.31 \\ \times \quad 4 \\ \hline \end{array}$$

2. Multiply. Rename.

$$\begin{array}{r} 1 \\ \$4.31 \\ \times \quad 4 \\ \hline 17\ 24 \end{array}$$

3. Put a decimal point and a dollar sign in the answer.

$$\begin{array}{r} 1 \\ \$4.31 \\ \times \quad 4 \\ \hline \$17.24 \end{array} \longleftarrow \text{two decimal places}$$

Multiply.

1.

$$\$2.97 \times 5 =$$

$$\begin{array}{r} 4\ 3 \\ \$2.97 \\ \times \quad 5 \\ \hline \$14.85 \end{array}$$

2.

$$\$9.49 \times 6 =$$

3.

$$\$3.26 \times 9 =$$

4.

$$\$7.85 \times 8 =$$

5.

$$\begin{array}{r} \$1.99 \\ \times \quad 7 \\ \hline \end{array}$$

6.

$$\begin{array}{r} \$6.25 \\ \times \quad 4 \\ \hline \end{array}$$

7.

$$\begin{array}{r} \$8.03 \\ \times \quad 6 \\ \hline \end{array}$$

8.

$$\begin{array}{r} \$9.60 \\ \times \quad 5 \\ \hline \end{array}$$

9.

$$\$ .57 \times 9 =$$

$$\begin{array}{r} 6 \\ \$ .57 \\ \times \quad 9 \\ \hline \$5.13 \end{array}$$

10.

$$\$ .08 \times 2 =$$

11.

$$\$ .30 \times 7 =$$

12.

$$\$ .99 \times 3 =$$

Multiplying Money

You multiply larger amounts of money the same way you multiply smaller amounts. Be sure to include a decimal point and dollar sign in the answer.

Use These Steps

Multiply $10.92 × 11

1. Set up the problem.

$$\begin{array}{r} \$10.92 \\ \times \quad 11 \\ \end{array}$$

2. Multiply.

$$\begin{array}{r} \$10.92 \\ \times \quad 11 \\ \hline 1092 \\ + \; 1092 \\ \hline 12012 \\ \end{array}$$

3. Put a decimal point and dollar sign in the answer.

$$\begin{array}{r} \$\ 10.92 \\ \times \quad 11 \\ \hline 10\ 92 \\ + \; 109\ 2 \\ \hline \$120.12 \\ \end{array}$$ ◄——— two decimal places

Multiply.

1.
$9.52 × 25 =

$$\begin{array}{r} \$9.52 \\ \times \quad 25 \\ \hline 47\ 60 \\ + \; 190\ 4 \\ \hline \$238.00 \\ \end{array}$$

2. $13.79 × 32 =

3. $25.39 × 19 =

4. $49.97 × 12 =

5.
$$\begin{array}{r} \$79.38 \\ \times \quad 24 \\ \end{array}$$

6.
$$\begin{array}{r} \$102.45 \\ \times \quad 52 \\ \end{array}$$

7.
$$\begin{array}{r} \$98.93 \\ \times \quad 17 \\ \end{array}$$

8.
$$\begin{array}{r} \$306.25 \\ \times \quad 44 \\ \end{array}$$

9. Adela pays $325.50 for rent each month. How much rent does she pay a year?
(Hint: 1 year = 12 months)

10. Zia pays $143.27 for her car payment each month. If she pays this amount for 48 months, how much will she pay in all?

Answer _____

Answer _____

Multiplying Money by 10, 100, and 1,000

When you multiply by 10, 100, or 1,000, you don't have to write a row of partial products with zeros. There is an easier way to multiply amounts of money by 10, 100, or 1,000.

Use These Steps

Multiply $9.98 × 100

1. Since there are two zeros in 100, add two zeros to $9.98.

   $9.98 × 100 = $9.9800

2. Move the decimal point two places to the right. Be sure a dollar sign is in the answer.

   $9.9800 = $998.00

Multiply.

1. $ .49 × 10 = $4.90

2. $ .25 × 10 =

3. $ .90 × 10 =

4. $2.31 × 10 =

5. $3.90 × 10 =

6. $58.06 × 10 =

7. $5.57 × 100 = $557.00

8. $13.20 × 100 =

9. $76.09 × 100 =

10. $240.65 × 100 =

11. $653.30 × 100 =

12. $500.03 × 100 =

13. $7.92 × 1,000 = $7,920.00

14. $1.40 × 1,000 =

15. $4.06 × 1,000 =

16. $25.29 × 1,000 =

17. $80.30 × 1,000 =

18. $320.07 × 1,000 =

19. $412.27 × 10 =

20. $ .67 × 100 =

21. $37.04 × 1,000 =

Multiplying Decimals by Whole Numbers

To multiply other decimals by whole numbers, be sure to line up the
digits the same way you do with amounts of money. The decimal point
in the answer needs to show the total number of decimal places to the
right of the decimal point in the problem.

$$4.2 \times 3 =$$

$$\begin{array}{r} 4.2 \\ \times\ \ 3 \\ \hline 12.6 \end{array} \leftarrow \text{one decimal place} \\ \leftarrow \text{no decimal places} \\ \leftarrow \text{one decimal place}$$

Use These Steps

Multiply 9.62 × 5

1. Set up the problem.

$$\begin{array}{r} 9.62 \\ \times\ \ \ 5 \\ \hline \end{array}$$

2. Multiply. Rename.

$$\begin{array}{r} {\scriptstyle 3\,1} \\ 9.62 \\ \times\ \ \ 5 \\ \hline 4810 \end{array}$$

3. Put a decimal point in the answer.

$$\begin{array}{r} {\scriptstyle 3\,1} \\ 9.62 \\ \times\ \ \ 5 \\ \hline 48.10 \end{array} \leftarrow \text{two decimal places} \\ \leftarrow \text{no decimal places} \\ \leftarrow \text{two decimal places}$$

Multiply.

1.

$$3.6 \times 9 =$$

$$\begin{array}{r} {\scriptstyle 5} \\ 3.6 \\ \times\ \ 9 \\ \hline 32.4 \end{array}$$

2. $.5 \times 4 =$

3. $7.2 \times 8 =$

4. $16.4 \times 7 =$

5.

$$.32 \times 5 =$$

$$\begin{array}{r} {\scriptstyle 1} \\ .32 \\ \times\ \ 5 \\ \hline 1.60 \end{array}$$

6. $10.51 \times 6 =$

7. $.08 \times 3 =$

8. $2.78 \times 9 =$

9.

$$\begin{array}{r} 7.3 \\ \times\ \ 4 \\ \hline \end{array}$$

10.

$$\begin{array}{r} .06 \\ \times\ \ 7 \\ \hline \end{array}$$

11.

$$\begin{array}{r} .7 \\ \times\ 5 \\ \hline \end{array}$$

12.

$$\begin{array}{r} 30.29 \\ \times\ \ \ \ 2 \\ \hline \end{array}$$

13.

$$\begin{array}{r} 51.02 \\ \times\ \ \ \ 6 \\ \hline \end{array}$$

14.

$$\begin{array}{r} 34.8 \\ \times\ \ 2 \\ \hline \end{array}$$

15.

$$\begin{array}{r} .45 \\ \times\ \ 4 \\ \hline \end{array}$$

16.

$$\begin{array}{r} 10.01 \\ \times\ \ \ \ 3 \\ \hline \end{array}$$

Multiplying Decimals by Whole Numbers

Count the places to the right of the decimal point in the problem. Be sure the decimal point in the answer shows the total number of decimal places to the right of the decimal point in the problem.

Use These Steps

Multiply 5.039 × 17

1. Set up the problem.

$$\begin{array}{r} 5.039 \\ \times\ \ 17 \\ \hline \end{array}$$

2. Multiply. Rename.

$$\begin{array}{r} 26 \\ 5.039 \\ \times\ \ 17 \\ \hline 35273 \\ +\ 5039 \\ \hline 85663 \end{array}$$

3. Put a decimal point in the answer.

$$\begin{array}{r} 26 \\ 5.039 \\ \times\ \ 17 \\ \hline 35\ 273 \\ +\ 50\ 39 \\ \hline 85.663 \end{array}$$

5.039 ◄— three decimal places
17 ◄— no decimal places
85.663 ◄— three decimal places

Multiply.

1.
2.341 × 25 =

$$\begin{array}{r} 2.341 \\ \times\ \ 25 \\ \hline 11\ 705 \\ +\ 46\ 82 \\ \hline 58.525 \end{array}$$

2.
.053 × 31 =

3.
28.027 × 46 =

4.
39.005 × 53 =

5.
6.218 × 182 =

6.
.509 × 246 =

7.
12.098 × 425 =

8.
37.186 × 931 =

9. In one state you pay $.085 in sales tax for each dollar you spend. If you spend $14, how much tax will you pay? Round the answer to the nearest hundredth, or cent.
(Hint: Multiply the tax rate by the amount of the purchase.)

10. The sales tax rate in another state is $.065. If you spend $120, how much will the tax be? Round the answer to the nearest hundredth, or cent.

Answer_____

Answer_____

Multiplying Decimals by 10, 100, and 1,000

To multiply a decimal by 10, 100, or 1,000, move the decimal point to the right the same number of places as there are zeros in the number you are multiplying by. You may need to add one or more zeros to get the correct number of places.

$2.4 \times 10 = 2.4 = 24$ $2.4 \times 100 = 2.40 = 240$ $2.4 \times 1,000 = 2.400 = 2,400$

Use These Steps

Multiply $.913 \times 100$

1. Count how many zeros there are in 100.

   There are two.

2. Move the decimal point two places to the right.

   $.913 \times 100 = .913 = 91.3$

Multiply.

1. $.5 \times 10 = 5$

2. $3.2 \times 10 =$

3. $15.6 \times 10 =$

4. $132.9 \times 10 =$

5. $6.17 \times 10 = 61.7$

6. $.98 \times 10 =$

7. $72.36 \times 10 =$

8. $241.05 \times 10 =$

9. $1.3 \times 100 = 130$

10. $.5 \times 100 =$

11. $8.6 \times 100 =$

12. $510.7 \times 100 =$

13. $4.67 \times 100 =$

14. $1.58 \times 100 =$

15. $.98 \times 100 =$

16. $423.02 \times 100 =$

17. $.5 \times 1,000 = 500$

18. $19.3 \times 1,000 =$

19. $5.7 \times 1,000 =$

20. $115.2 \times 1,000 =$

21. $.32 \times 1,000 =$

22. $8.51 \times 1,000 =$

23. $1.07 \times 1,000 =$

24. $266.29 \times 1,000 =$

25. $.865 \times 10 =$

26. $1.207 \times 100 =$

27. $10.036 \times 1,000 =$

28. $372.331 \times 10 =$

29. $7.314 \times 1,000 =$

30. $.802 \times 10 =$

31. $11.006 \times 1,000 =$

32. $659.078 \times 100 =$

Multiplying Decimals by Decimals

To multiply a decimal by another decimal, first multiply the way you do whole numbers. Then count the number of decimal places in the problem. Put a decimal point in the answer to show the total number of decimal places you counted. Always count from the right.

$$
\begin{array}{r}
10.39 \leftarrow \text{two decimal places} \\
\times \quad .06 \leftarrow \text{two decimal places} \\
\hline
.6234 \leftarrow \text{four decimal places}
\end{array}
$$

Use These Steps

Multiply 3.247 × 1.5

1. Set up the problem.

$$
\begin{array}{r}
3.247 \\
\times \quad 1.5 \\
\end{array}
$$

2. Multiply. Rename.

$$
\begin{array}{r}
{\scriptstyle 1\ 23} \\
3.247 \\
\times \quad 1.5 \\
\hline
16235 \\
+\ 3247 \\
\hline
48705
\end{array}
$$

3. Put a decimal point in the answer.

$$
\begin{array}{r}
{\scriptstyle 1\ 23} \\
3.247 \leftarrow \text{three decimal places} \\
\times \quad 1.5 \leftarrow \text{one decimal place} \\
\hline
1\ 6235 \\
+\ 3\ 247 \\
\hline
4.8705 \leftarrow \text{four decimal places}
\end{array}
$$

Multiply.

1.
$5.6 \times 1.8 =$

$$
\begin{array}{r}
{\scriptstyle 4} \\
5.6 \\
\times 1.8 \\
\hline
4\ 4\ 8 \\
+\ 5\ 6 \\
\hline
10.0\ 8
\end{array}
$$

2.
$9.04 \times 3.2 =$

3.
$.99 \times .5 =$

4.
$.3 \times .7 =$

5.
$15.3 \times 3.12 =$

6.
$8.29 \times 4.33 =$

7.
$7.851 \times .03 =$

8.
$25.2 \times .75 =$

9.
$$
\begin{array}{r}
82.9 \\
\times \quad 2.6 \\
\hline
\end{array}
$$

10.
$$
\begin{array}{r}
60.08 \\
\times \quad .07 \\
\hline
\end{array}
$$

11.
$$
\begin{array}{r}
9.272 \\
\times \quad 5.01 \\
\hline
\end{array}
$$

12.
$$
\begin{array}{r}
75.2 \\
\times 3.29 \\
\hline
\end{array}
$$

Multiplying Decimals by Decimals

Sometimes the answer does not have enough digits to show the correct number of decimal places. When this happens, you need to insert one or more zeros between the answer and the decimal point. The zeros serve as place holders so you can show the correct number of places.

$$
\begin{array}{r}
.185 \leftarrow \text{three decimal places} \\
\times\ \ .12 \leftarrow \text{two decimal places} \\
\hline
370 \\
+\ \ 185 \\
\hline
.02220 \leftarrow \text{five decimal places} \\
\uparrow\kern-1em\underline{\qquad\qquad}\text{insert a zero}
\end{array}
$$

Use These Steps

Multiply .4 × .1

1. Set up the problem.

$$
\begin{array}{r}
.4 \\
\times .1 \\
\hline
\end{array}
$$

2. Multiply.

$$
\begin{array}{r}
.4 \\
\times .1 \\
\hline
4
\end{array}
$$

3. Insert a zero. Put a decimal point in the answer.

$$
\begin{array}{r}
.4 \leftarrow \text{one decimal place} \\
\times .1 \leftarrow \text{one decimal place} \\
\hline
.04 \leftarrow \text{two decimal places}
\end{array}
$$

insert a zero

Multiply.

1.
.3 × .2 =

$$
\begin{array}{r}
.3 \\
\times .2 \\
\hline
.06
\end{array}
$$

2.
.4 × .2 =

3.
.16 × .12 =

4.
.09 × .06 =

5.
41.3 × .001 =

6.
.303 × .22 =

7.
1.31 × .021 =

8.
.095 × .067 =

9.
5.15 × .3 =

10.
6.4 × .104 =

11.
9.08 × .13 =

12.
70.1 × .04 =

Real-Life Application

You multiply decimals to find the total
cost when you know the price per unit.

Example Honey costs $3.82 cents a
pound. How much does
five pounds cost?

$$
\begin{array}{r}
4\,1 \\
\$3.82 \\
\times \quad 5 \\
\hline
\$19.10
\end{array}
$$

Five pounds of honey costs $19.10.

Solve. Round the answers to the nearest hundredth, or cent.

1. Apples cost $ .69 a pound. How much
   do 10 pounds cost?

   Answer_____

2. Eggs cost $ .89 a dozen. How much do
   3 dozen cost?

   Answer_____

3. Swordfish costs $13.00 a pound. How
   much does .5 pound cost?

   Answer_____

4. Granola costs $2.99 a pound. How
   much does 1.5 pounds cost?

   Answer_____

5. Fancy coffee beans cost $5.75 a pound.
   How much do 2.35 pounds cost?

   Answer_____

6. One pound of butter costs $1.69. How
   much does .25 pound cost?

   Answer_____

Mixed Review

Add, subtract, or multiply.

1.
$3.95
+ .32

2.
$1.39
− .79

3.
$2.29
× 6

4.
$5.00
− 4.37

5.
$10.98
+ 3.65

6.
$ .59
× 12

7.
$36.00
− 9.23

8.
$8.99
× 10

9.
$ .25
+ .75

10.
$12.05
− 6.92

11.
.47
+ .93

12.
2.05
− .79

13.
31.3
× 4.9

14.
4.
+ 3.6

15.
6.01
− 4.99

16.
1.5
− .7

17.
14.06
× 20

18.
6.23
+ .59

19.
.4
× .2

20.
25.3
× 8

21.
.325
+ .987

22.
7.022
− 1.853

23.
10.344
+ 5.6

24.
1.472
× 10

25.
16.021
× .7

26.
13.491
+ .281

27.
532.065
− 93.279

28.
49.006
× .32

29.
1.610
− .234

30.
.256
× .19

31.
8.521
× 25

32.
2.035
− .216

33.
154.
− 29.326

34.
1.846
× .75

35.
.051
× .09

Dividing Money

To divide amounts of money, first set up the problem. Put a decimal point in the answer above the decimal point in the problem. Put a dollar sign in the answer.

$$\$7.70 \div 5$$

$$\begin{array}{r} \$1.54 \\ 5\overline{)\$7.70} \\ -5 \\ \hline 2\,7 \\ -2\,5 \\ \hline 20 \\ -20 \\ \hline 0 \end{array}$$

Use These Steps

Divide $4.96 ÷ 8

1. Set up the problem.

$$8\overline{)\$4.96}$$

2. Put a dollar sign and a decimal point in the answer.

$$\begin{array}{r} \$. \\ 8\overline{)\$4.96} \end{array}$$

3. Divide.

$$\begin{array}{r} \$.62 \\ 8\overline{)\$4.96} \\ -4\,8 \\ \hline 16 \\ -16 \\ \hline 0 \end{array}$$

Divide.

1.
$$\$5.04 \div 6 =$$

$$\begin{array}{r} \$.84 \\ 6\overline{)\$5.04} \\ -4\,8 \\ \hline 24 \\ -24 \\ \hline 0 \end{array}$$

2.
$$\$2.79 \div 9 =$$

3.
$$\$.36 \div 4 =$$

4.
$$\$1.00 \div 5 =$$

5.
$$\$9.84 \div 8 =$$

6.
$$\$12.60 \div 6 =$$

7.
$$\$17.37 \div 3 =$$

8.
$$\$62.86 \div 7 =$$

Dividing Money

Sometimes when you divide amounts of money, you have a remainder. When this happens, divide to three decimal places. Then round to the nearest hundredth. This is the same as rounding to the nearest cent.

Use These Steps

Divide $9.23 ÷ 16

1. Set up the problem. Put a dollar sign and a decimal point in the answer.

$$\begin{array}{r} \$\ .\underline{} \\ 16\overline{)\$9.23} \end{array}$$

2. Add a zero so that you can divide to three decimal places.

$$\begin{array}{r} \$\ .576 \\ 16\overline{)\$9.230} \\ -8\ 0 \\ \hline 1\ 23 \\ -1\ 12 \\ \hline 110 \\ -\ 96 \\ \hline 14 \end{array}$$

3. Ignore the remainder. Round the answer to the nearest hundredth, or cent.

$ .576 rounds to $ .58

Divide. Round each answer to the nearest hundredth, or cent.

1.
$$\$13.40 \div 32 =$$
$$\begin{array}{r} \$\ .418 \\ 32\overline{)\$13.400} \\ -12\ 8 \\ \hline 60 \\ -32 \\ \hline 280 \\ -256 \\ \hline 24 \end{array}$$
$ .418 rounds to $ .42

2.
$$\$21.36 \div 15 =$$

3.
$$\$427.00 \div 26 =$$

4.
$$\$5.89 \div 11 =$$

5. $49\overline{)\$178.90}$

6. $62\overline{)\$308.00}$

7. $36\overline{)\$27.06}$

8. $50\overline{)\$152.75}$

Dividing Money by 10, 100, and 1,000

When you divide amounts of money by 10, 100, or 1,000, you don't have to set up the problem. Move the decimal point to the left the same number of places as there are zeros in the number you are dividing by. You may need to insert zeros as place holders between the answer and the decimal point.

Use These Steps

Divide $ .92 ÷ 10

1. Since there is one zero in 10, move the decimal point one place to the left.

   $ .92 ÷ 10 = $ .92 = $ .092

   └── insert a zero

2. Round to two decimal places.

   $ .092 rounds to $ .09

Divide. Round to the nearest hundredth, or cent.

1.
$5.67 ÷ 10 = $ 5.67
$ .567 rounds to $ .57

2.
$ .40 ÷ 10 =

3.
$18.71 ÷ 10 =

4.
$4.01 ÷ 10 =

5.
$41.83 ÷ 10 =

6.
$ .78 ÷ 10 =

7.
$3.35 ÷ 100 = $ 3.35
$ .0335 rounds to $ .03

8.
$9.89 ÷ 100 =

9.
$10.06 ÷ 100 =

10.
$.65 ÷ 100 =

11.
$21.80 ÷ 100 =

12.
$127.35 ÷ 100 =

13.
$497.62 ÷ 1,000 = $ 497.62
$.49762 rounds to $.50

14.
$3,485.99 ÷ 1,000 =

15.
$700.00 ÷ 1,000 =

16.
$9.74 ÷ 1,000 =

17.
$35.75 ÷ 1,000 =

18.
$516.03 ÷ 1,000 =

19.
$30.77 ÷ 10 =

20.
$4.13 ÷ 100 =

21.
$16.50 ÷ 1,000 =

Dividing Decimals by Whole Numbers

To divide other decimals by whole numbers, set up the problem. Put a decimal point in the answer above the decimal point in the problem.

Use These Steps

Divide .408 ÷ 4

1. Set up the problem.

$$4\overline{)\,.408}$$

2. Put a decimal point in the answer.

$$4\overline{)\,.408}^{\,.}$$

3. Divide.

$$\begin{array}{r} .102 \\ 4\overline{)\,.408} \\ -\underline{4} \\ 00 \\ -\underline{0} \\ 08 \\ -\underline{8} \\ 0 \end{array}$$

Divide.

1.
$$7.36 \div 8 =$$
$$\begin{array}{r} .92 \\ 8\overline{)\,7.36} \\ -\underline{7\,2} \\ 16 \\ -\underline{16} \\ 0 \end{array}$$

2.
$$5.2 \div 4 =$$

3.
$$8.5 \div 5 =$$

4.
$$.627 \div 3 =$$

5.
$$.74 \div 2 =$$

6.
$$123.6 \div 6 =$$

7.
$$25.74 \div 9 =$$

8.
$$.861 \div 7 =$$

9.
$$10.2 \div 3 =$$

10.
$$3.005 \div 5 =$$

11.
$$1.68 \div 8 =$$

12.
$$9.2 \div 4 =$$

 62

Dividing Decimals by Whole Numbers

Sometimes when dividing decimals, you get a remainder. When this happens, you can continue to divide by adding zeros to the number you are dividing into. Stop dividing when the remainder is zero, or when there are enough decimal places to round to a certain place. For example, to round to tenths, divide to two decimal places.

Use These Steps

Divide 17.3 ÷ 6. Round to the nearest tenth.

1. Set up the problem. Put a decimal point in the answer.

$$6\overline{)17.3}$$

2. Add a zero so that you can divide to two decimal places.

$$\begin{array}{r} 2.88 \\ 6\overline{)17.30} \\ -\,12 \\ \hline 5\,3 \\ -\,4\,8 \\ \hline 50 \\ -\,48 \\ \hline 2 \end{array}$$

3. Round to the nearest tenth.

2.88 rounds to 2.9

Divide. Round to the nearest tenth.

1.
$$20.3 \div 3 =$$

$$\begin{array}{r} 6.76 \\ 3\overline{)20.30} \\ -\,18 \\ \hline 2\,3 \\ -\,2\,1 \\ \hline 20 \\ -\,18 \\ \hline 2 \end{array}$$

6.7 rounds to 6.8

2.
$$76.41 \div 8 =$$

3.
$$100.0 \div 7 =$$

4.
$$9.7 \div 9 =$$

Divide. Round to the nearest hundredth.

5.
$$1.2 \div 25 =$$

$$\begin{array}{r} .048 \\ 25\overline{)1.200} \\ -\,1\,00 \\ \hline 200 \\ -\,200 \\ \hline 0 \end{array}$$

.048 rounds to .05

6.
$$1{,}480.29 \div 37 =$$

7.
$$5.281 \div 40 =$$

8.
$$.99 \div 15 =$$

Dividing Decimals by 10, 100, and 1,000

When you divide by 10, 100, or 1,000, move the decimal point in the answer to the left the same number of places as there are zeros in the number you are dividing by.

Use These Steps

Divide 1.4 ÷ 100

1. There are two zeros in 100.

    100

2. Move the decimal point two places to the left. Insert a zero.

    1.4 ÷ 100 = 1.4 = .014

    └── insert a zero

Divide.

1.
.3 ÷ 10 = .03

2.
5.4 ÷ 10 =

3.
22.7 ÷ 10 =

4.
.72 ÷ 10 =

5.
45.36 ÷ 10 =

6.
298.54 ÷ 10 =

7.
4.5 ÷ 100 = .045

8.
.2 ÷ 100 =

9.
11.6 ÷ 100 =

10.
.86 ÷ 100 =

11.
3.09 ÷ 100 =

12.
247.41 ÷ 100 =

13.
92.3 ÷ 1,000 = .0923

14.
.4 ÷ 1,000 =

15.
311.6 ÷ 1,000 =

16.
.47 ÷ 1,000 =

17.
3.21 ÷ 1,000 =

18.
86.55 ÷ 1,000 =

19.
1.802 ÷ 100 =

20.
29.004 ÷ 1,000 =

21.
.407 ÷ 10 =

22.
100.255 ÷ 1,000 =

23.
.573 ÷ 10 =

24.
6.102 ÷ 100 =

25.
39.027 ÷ 100 =

26.
766.341 ÷ 10 =

27.
4.011 ÷ 1,000 =

28.
75.403 ÷ 100 =

29.
.886 ÷ 10 =

30.
137.429 ÷ 100 =

Dividing Whole Numbers by Decimals

To divide a whole number by a decimal, first move the decimal point to the right to make a whole number. Then move the decimal point in the number you are dividing into the same number of places to the right. Add one zero for each place you moved the decimal point. Put the decimal point in the answer directly above the decimal point in the number you are dividing into.

$$36 \div 1.5$$

$$
\begin{array}{r}
24. \\
1.5\overline{)360} \\
-30 \\
\hline
60 \\
-60 \\
\hline
0
\end{array}
$$

Use These Steps

Divide 48 ÷ .24

1. Set up the problem.

$$.24\overline{)48}$$

2. Move both decimal points two places to the right. Add two zeros. Put a decimal point in the answer.

$$.24\,\overline{)48.00}$$

3. Divide.

$$
\begin{array}{r}
200. \\
24\overline{)4800} \\
-48 \\
\hline
00 \\
-0 \\
\hline
00 \\
-0 \\
\hline
0
\end{array}
$$

Divide.

1.
$$69 \div 2.3 =$$

$$
\begin{array}{r}
30. \\
2.3\overline{)69.0} \\
-69 \\
\hline
00 \\
-0 \\
\hline
0
\end{array}
$$

2.
$$42 \div 3.5 =$$

3.
$$993 \div .3 =$$

4.
$$124 \div .4 =$$

5.
$$580 \div 7.25 =$$

6.
$$165 \div .55 =$$

7.
$$806 \div 8.06 =$$

8.
$$278 \div 1.39 =$$

Dividing Whole Numbers by Decimals

Sometimes when you divide whole numbers by decimals you get a remainder. When this happens, you can continue to divide by adding zeros to the number you are dividing into. Stop dividing when the remainder is zero, or when there are enough decimal places in the answer to round to a certain place.

Use These Steps

Divide 3 ÷ .8 Round to the nearest tenth.

1. Set up the problem.

$$.8\overline{)3}$$

2. Move both decimal points one place to the right. Add a zero. Put a decimal point in the answer.

$$.8\overline{)3.0}$$

3. Divide. Add zeros. Round to the nearest tenth.

```
        3.75
  8)30.00
   − 24
      6 0
    − 5 6
        40
      − 40
         0
```
3.75 rounds to 3.8

Divide. Round to the nearest tenth.

1.
$$6 ÷ .7 =$$
```
       8.57
  .7)60.00
   − 56
      4 0
    − 3 5
        50
      − 49
         1
```
8.57 rounds to 8.6

2.
$$10 ÷ 1.3 =$$

3.
$$12 ÷ 4.1 =$$

4.
$$4 ÷ 2.2 =$$

Divide. Round to the nearest hundredth.

5.
$$50 ÷ .75 =$$

6.
$$100 ÷ 6.49 =$$

7.
$$29 ÷ 10.8 =$$

8.
$$76 ÷ 4.3 =$$

Dividing Decimals by Decimals

To divide a decimal by another decimal, you need to move the decimal points in both numbers the same number of places to the right. You may need to add zeros to the number you are dividing into.

Use These Steps

Divide 5.25 ÷ .75

1. Set up the problem.

$$.75\overline{)5.25}$$

2. Move both decimal points two places to the right. Put a decimal point in the answer.

$$.75\overline{)5.25}$$

3. Divide.

$$\begin{array}{r} 7. \\ 75\overline{)525} \\ -525 \\ \hline 0 \end{array}$$

Divide.

1.
$$.91 \div .7 =$$

$$\begin{array}{r} 1.3 \\ .7\overline{).91} \\ -7 \\ \hline 21 \\ 21 \\ \hline 0 \end{array}$$

2.
$$5.4 \div .3 =$$

3.
$$8.2 \div 4.1 =$$

4.
$$26.4 \div .24 =$$

5.
$$.36 \div .6 =$$

6.
$$9.7 \div .5 =$$

7.
$$29.43 \div 2.7 =$$

8.
$$.76 \div 1.9 =$$

9.
$$.22\overline{).352}$$

10.
$$3.4\overline{)21.42}$$

11.
$$.15\overline{)4.83}$$

12.
$$.72\overline{)1.512}$$

Dividing Decimals by Decimals

Remember to move both decimal points the same number of places before dividing.

Use These Steps

Divide $2.3 \div .6$ **Round to the nearest hundredth.**

1. Set up the problem.

$$.6\overline{)2.3}$$

2. Move both decimal points one place to the right. Put a decimal point in the answer.

$$.6\overline{)2.3}$$

3. Divide. Add zeros. Round to the nearest hundredth.

```
      3.833
  6)23.000
   - 18
     5 0
    - 4 8
      20
     - 18
       20
      - 18
        2
```

3.833 rounds to 3.83

Divide. Round to the nearest tenth.

1.
$$.75 \div 1.3 =$$

```
      .57
1.3 ).750
    - 65
     100
    -  91
       9
```
.57 rounds to .6

2.
$$.348 \div .9 =$$

3.
$$18.5 \div .239 =$$

4.
$$2.61 \div .4 =$$

Divide. Round to the nearest hundredth.

5.
$$67.6 \div 3.22 =$$

6.
$$.881 \div .25 =$$

7.
$$7.47 \div 2.4 =$$

8.
$$3.143 \div .5 =$$

Problem Solving: Using Estimating

Often you do not need to find the exact answer to a problem. Estimating will give you an answer that is close to the exact answer. You are estimating when you round a decimal to the nearest whole number.

Example   On a week-long camping trip, 8 people hiked a total of 31.62 miles in one day. About how many miles on the average did each person hike?

▶ **Step 1.**   Estimate by rounding 31.62 to the nearest whole number.

   31.62 rounds to 32

▶ **Step 2.**   Divide.

$$
\begin{array}{r}
4 \\
8\overline{)32} \\
-32 \\
\hline
0
\end{array}
$$

Each person hiked about 4 miles.

Round each amount to the nearest whole number. Then solve.

1. On Saturday and Sunday, the campers went on a 135.5-mile canoe trip. About how many miles did they average each day?

    Answer_____

2. One day six of the campers went horseback riding. They rode 17.9 miles in 6.2 hours. About how many miles per hour did they travel?

    Answer_____

3. The 6 riders carried 96.1 pounds of supplies. If they divided the supplies equally, about how many pounds did each rider carry?

    Answer_____

4. To get to the campground, Katie drove 249.82 miles. She made the trip in 5.2 hours. About how many miles per hour did she travel?

    Answer_____

5. Katie used 9.98 gallons of gas to drive 249.82 miles. About how many miles does her car travel on a gallon of gas?

Answer_____

6. Katie's total expenses for the trip were $328.60. On the average, how much did she spend each day for the 7 days?

Answer_____

7. One night the campers ate 7.6 pounds of barbequed chicken. About how much chicken did each of the 8 campers eat if they divided it evenly?

Answer_____

8. Beef ribs cost the campers a total of $47.96 for 16 pounds. About what was the cost per pound?

Answer_____

9. The 8 campers split their expenses evenly. They spent $560.07 on food. About how much did each person pay for food?

Answer_____

10. If the campers spent $692.90 for equipment rental and supplies for the week, about how much did the group spend each day?
(Hint: There are 7 days in a week.)

Answer_____

Unit 3 *Review*

Multiply.

1.
$\$ .67 \times 4 =$

2.
$\$ .99 \times 6 =$

3.
$\$7.49 \times 5 =$

4.
$\$1.35 \times 8 =$

5.
$\$11.87 \times 12 =$

6.
$\$69.98 \times 24 =$

7.
$\$ .50 \times 100 =$

8.
$\$2.95 \times 1,000 =$

9.
$.8 \times 6 =$

10.
$3.2 \times 7 =$

11.
$.06 \times 9 =$

12.
$15.29 \times 5 =$

13.
$1.201 \times 25 =$

14.
$.851 \times 14 =$

15.
$21.03 \times 10 =$

16.
$.34 \times 1,000 =$

17.
$3.2 \times 1.4 =$

18.
$10.05 \times .33 =$

19.
$.7 \times .9 =$

20.
$12.3 \times 4.27 =$

21.
$.2 \times .1 =$

22.
$.05 \times .3 =$

23.
$.3 \times .9 =$

24.
$2.3 \times .08 =$

Divide.

25.
$\$9.20 \div 4 =$

26.
$\$14.00 \div 8 =$

27.
$\$ .95 \div 5 =$

28.
$\$46.92 \div 17 =$

Divide. Round each answer to the nearest hundredth.

29.
$200.00 ÷ 10 =

30.
$ .98 ÷ 10 =

31.
$3.47 ÷ 100 =

32.
$92.06 ÷ 1,000 =

33.
.40 ÷ 8 =

34.
2.42 ÷ 11 =

35.
.379 ÷ 42 =

36.
31.84 ÷ 6 =

37.
2.1 ÷ 10 =

38.
17.37 ÷ 100 =

39.
416.9 ÷ 100 =

40.
.2 ÷ 1,000 =

41.
124 ÷ .4 =

42.
28 ÷ .36 =

43.
2.9 ÷ 1.4 =

44.
8.906 ÷ .15 =

Below is a list of the problems in this review and the pages on which the skills are taught. If you missed any problems, turn to the pages listed and practice the skills. Then correct the problems you missed in the Unit Review.

Problems	Pages	Problems	Pages
1-8	49-51	33-36	62-63
9-16	52-54	37-40	64
17-24	55-56	41-42	65-66
25-32	59-61	43-44	67-68

Unit 4 RATIOS, PROPORTIONS, AND PERCENTS

You must understand the meaning of ratios, proportions, and percents before you can work with percent problems. Percents, like fractions and decimals, are used to show part of a whole. For example, if you have $20 and you spend $10, you have spent $\frac{10}{20}$ or 50% of your money.

In this unit, you will learn how to write ratios, proportions, and percents. You will learn how to change from fractions to decimals and to percents, and how to change from percents to decimals and to fractions.

Getting Ready

You should be familiar with the skills on this page and the next before you begin this unit. To check your answers, turn to page 177.

▶ You will need to use place value when working with percents.

Write the place value of the underlined digit in each number.

1.	13.<u>2</u> two tenths	**2.**	1.32<u>4</u>5 _____
3.	.0<u>7</u> _____	**4.**	29.03<u>6</u> _____
5.	9.0<u>0</u>8 _____	**6.**	6.<u>9</u>8 _____

Getting Ready

To change a fraction to a decimal, divide the numerator by the denominator. Divide until there is no remainder.

Change each fraction to a decimal.

7.

$\frac{1}{2} = .5$

$$\begin{array}{r} .5 \\ 2{\overline{)1.0}} \\ -10 \\ \hline 0 \end{array}$$

8.

$\frac{3}{4} =$

9.

$\frac{2}{5} =$

10.

$\frac{3}{10} =$

11.

$\frac{1}{20} =$

12.

$\frac{3}{5} =$

13.

$\frac{9}{20} =$

14.

$\frac{4}{25} =$

For review, see Unit 1, page 21.

When changing some fractions to decimals, your answer may have a remainder.

Change each fraction to a decimal with two digits to the right of the decimal point. Write the remainder as a fraction.

15.

$\frac{1}{3} = .33\frac{1}{3}$

$$\begin{array}{r} .33\frac{1}{3} \\ 3{\overline{)1.00}} \\ -9 \\ \hline 10 \\ -9 \\ \hline 1 \end{array}$$

16.

$\frac{5}{6} =$

17.

$\frac{4}{9} =$

18.

$\frac{2}{3} =$

19.

$\frac{7}{15} =$

20.

$\frac{5}{18} =$

21.

$\frac{2}{11} =$

22.

$\frac{3}{7} =$

For review, see Unit 1, page 22.

Ratios

A ratio is a fraction that shows a relationship between two numbers. For example, if there are 4 cashiers and 27 customers, then the ratio of cashiers to customers can be shown as the fraction $\frac{4}{27}$.

Always write the first number in the relationship as the numerator and the second number as the denominator.

Look at the examples below. Notice that the denominators may be smaller than the numerators.

5 hits out of 9 times at bat = $\frac{5}{9}$ 70 pitches in 6 innings = $\frac{70}{6}$ 1 win and 1 loss = $\frac{1}{1}$

A ratio may be reduced without changing the relationship. It can have a denominator of 1, but it cannot be changed to a mixed number or a whole number.

Use These Steps

Write a ratio to show a car traveling 48 miles on 3 gallons of gas.

1. Write the first number in the relationship as the numerator of the fraction.

$$\frac{48}{}$$

2. Write the second number as the denominator.

$$\frac{48}{3}$$

3. Reduce.

$$\frac{48 \div 3}{3 \div 3} = \frac{16}{1}$$

Write a ratio for each problem. Reduce if possible.

1.
   7 out of 10 houses
   $$\frac{7}{10}$$

2.
   6 sodas for 9 people

3.
   99 out of 100 doctors

4.
   50 yards in 10 seconds
   $$\frac{50}{10} = \frac{50 \div 10}{10 \div 10} = \frac{5}{1}$$

5.
   2 pain relievers in 1 pill

6.
   5 fingers on 1 hand

7.
   6 muffins for 6 people

8.
   4 wins and 4 losses

9.
   2 out of 3 dentists

10. Richard worked 8 hours on Friday and 5 hours on Saturday. What is the ratio of the number of hours he worked on Friday to the hours he worked on Saturday?

11. Sara prepared 200 dinners for a party she catered. Only 194 guests went to the party. Write the ratio of dinners she prepared to the number of guests at the party.

Answer_____

Answer_____

Equal Ratios

You can change ratios to higher terms by multiplying both the numerator and the denominator by the same number. For example, if you can buy 3 oranges for $1, $\frac{3}{1}$, you can buy 6 oranges for $2, $\frac{6}{2}$, because $\frac{3}{1} = \frac{6}{2}$.

You can use cross-multiplication to determine if two ratios are equal. To cross-multiply, multiply the numbers in the opposite corners. If the ratios are equal, the answers to the cross-multiplication will be equal.

Use These Steps

Change $\frac{7}{2}$ to an equal ratio with 6 as the denominator.

1. Raise $\frac{7}{2}$ to higher terms with a denominator of 6.

$$\frac{7}{2} = \frac{7 \times 3}{2 \times 3} = \frac{21}{6}$$

2. Check by cross-multiplication.

$$\frac{7}{2} \diagup \diagdown \frac{21}{6}$$

$$7 \times 6 = 2 \times 21$$
$$42 = 42$$

Change to higher terms. Check by cross-multiplication.

1.
$$\frac{3}{8} = \frac{\boxed{6}}{16}$$
$$\frac{3}{8} = \frac{3 \times 2}{8 \times 2} = \frac{6}{16}$$
Check:
$$\frac{3}{8} \diagup \diagdown \frac{6}{16}$$
$$3 \times 16 = 8 \times 6$$
$$48 = 48$$

2.
$$\frac{4}{9} = \frac{\square}{27}$$

3.
$$\frac{7}{10} = \frac{\square}{20}$$

4.
$$\frac{5}{6} = \frac{\square}{36}$$

5.
$$\frac{1}{2} = \frac{2}{\square}$$

6.
$$\frac{2}{3} = \frac{6}{\square}$$

7.
$$\frac{4}{7} = \frac{8}{\square}$$

8.
$$\frac{3}{5} = \frac{9}{\square}$$

9. The shipping department at Hartman's can process 2 orders in 5 minutes. How much time do they need to process 16 orders?

10. If the computer prints 2 purchase orders in 3 minutes, how many orders can it print in 15 minutes?

Answer_____

Answer_____

Equal Ratios

Equal ratios are equal fractions. If the numerator or the denominator is missing from one fraction, you can find it by multiplying (changing to higher terms) or dividing (reducing to lower terms). Write n to show a missing number.

$$\frac{n}{2} = \frac{1}{2}$$

Use These Steps

Find n. $\frac{n}{12} = \frac{2}{3}$

1. To find the missing number, n, change $\frac{2}{3}$ to an equal ratio with a denominator of 12.

$$\frac{n}{12} = \frac{2}{3}$$

$$\frac{2}{3} = \frac{2 \times 4}{3 \times 4} = \frac{8}{12}$$

$$n = 8$$

2. Check by cross-multiplication.

  $\frac{8}{12} \quad \frac{2}{3}$

$$8 \times 3 = 12 \times 2$$
$$24 = 24$$

Find n. Check by cross-multiplication.

1.
$$\frac{2}{6} = \frac{n}{3}$$
$$\frac{2}{6} = \frac{2 \div 2}{6 \div 2} = \frac{1}{3}$$
$$n = 1$$
Check:
$$\frac{2}{6} \quad \frac{1}{3}$$
$$2 \times 3 = 6 \times 1$$
$$6 = 6$$

2.
$$\frac{n}{20} = \frac{4}{5}$$

3.
$$\frac{9}{21} = \frac{3}{n}$$

4.
$$\frac{n}{8} = \frac{3}{4}$$

5.
$$\frac{n}{4} = \frac{3}{12}$$

6.
$$\frac{5}{n} = \frac{10}{12}$$

7.
$$\frac{4}{14} = \frac{n}{7}$$

8.
$$\frac{4}{n} = \frac{20}{25}$$

9. Darryl can rent 3 videos for $6. How many videos can he rent for $2?

10. Darryl can rent a VCR for 2 days for $5. How much does it cost to rent a VCR for 4 days?

Answer_____

Answer_____

Proportions

Equal ratios are called proportions. Proportions are used to solve problems when the relationship between two things does not change. For example, if you know that your car uses 1 gallon of gas to go 24 miles, you can find out how many gallons of gas you need to go 240 miles by writing a proportion and cross multiplying to find the missing number.

$$\frac{1 \text{ gallon}}{24 \text{ miles}} = \frac{n \text{ gallons}}{240 \text{ miles}}$$

When you cross-multiply, first multiply n by the number in the opposite corner. Then write the answer on the left side of the equal sign.

$$24 \times n = 1 \times 240$$

Use These Steps

Find n. $\frac{1}{2} = \frac{n}{24}$

1. Cross-multiply. 2n means 2 times n.

$$\frac{1}{2} \diagup\hspace{-0.5em}\diagdown \frac{n}{24}$$

$$2 \times n = 1 \times 24$$
$$2n = 24$$

2. Divide the number on the right of the equal sign by the number next to n, 2.

$$n = 24 \div 2 = 12$$

3. Check by substituting the answer, 12, for n. Cross-multiply.

$$\frac{1}{2} \diagup\hspace{-0.5em}\diagdown \frac{12}{24}$$

$$2 \times 12 = 1 \times 24$$
$$24 = 24$$

Find n. Check by substituting the answer and cross-multiplying.

1.
$$\frac{n}{5} \diagup\hspace{-0.5em}\diagdown \frac{6}{15}$$
$$n \times 15 = 5 \times 6$$
$$15n = 30$$
$$n = 30 \div 15 = 2$$
Check:
$$\frac{2}{5} \diagup\hspace{-0.5em}\diagdown \frac{6}{15}$$
$$2 \times 15 = 5 \times 6$$
$$30 = 30$$

2. $\frac{3}{4} = \frac{12}{n}$

3. $\frac{n}{9} = \frac{1}{3}$

4. $\frac{3}{n} = \frac{15}{25}$

5. $\frac{3}{24} = \frac{n}{8}$

6. $\frac{n}{32} = \frac{15}{16}$

7. $\frac{7}{10} = \frac{21}{n}$

8. $\frac{n}{4} = \frac{25}{100}$

Real-Life Application

Proportions are especially useful for solving problems that you cannot do easily in your head.

Example If 2 grapefruit cost $ .75, how much do 3 grapefruit cost?

$$\frac{2 \text{ grapefruit}}{\$ .75} = \frac{3 \text{ grapefruit}}{n}$$

$$2 \times n = \$.75 \times 3$$
$$2n = \$2.25$$
$$n = \$2.25 \div 2 = \$1.125$$

Grapefruit
2 for $.75

Round the answer to the nearest hundredth, or cent.

$1.125 rounds to $1.13, so 3 grapefruit cost $1.13.

Write a proportion. Solve. Round the answers to the nearest hundredth, or cent.

1. If cucumbers are 2 for $ .75, how much do 9 cucumbers cost?

2. If bananas are 3 pounds for $1.25, how much do 4 pounds cost?

Answer_____

Answer_____

3. If tomatoes are 5 pounds for $3.00, how much do 2 pounds cost?

4. If lemons are 3 for $1.00, how much do 10 lemons cost?

Answer_____

Answer_____

Mixed Review

Write a ratio for each problem. Reduce if possible.

1. 2 coupons to a customer

2. 18 socks to 24 shoes

3. 4 chances out of 100

4. 3 pounds for $1

5. 85 miles in 2 hours

6. 10 trucks to 10 drivers

Find n by reducing or changing to higher terms. Check by cross-multiplying.

7. $\dfrac{1}{3} = \dfrac{n}{6}$

8. $\dfrac{3}{27} = \dfrac{n}{9}$

9. $\dfrac{8}{10} = \dfrac{4}{n}$

10. $\dfrac{5}{6} = \dfrac{30}{n}$

11. $\dfrac{n}{40} = \dfrac{3}{8}$

12. $\dfrac{n}{7} = \dfrac{21}{49}$

13. $\dfrac{16}{n} = \dfrac{4}{5}$

14. $\dfrac{3}{n} = \dfrac{12}{100}$

Find n by cross-multiplying. Check by substituting the answer and cross-multiplying.

15. $\dfrac{1}{4} = \dfrac{n}{24}$

16. $\dfrac{9}{n} = \dfrac{3}{20}$

17. $\dfrac{n}{10} = \dfrac{2}{5}$

18. $\dfrac{4}{n} = \dfrac{2}{7}$

19. $\dfrac{n}{10} = \dfrac{1}{2}$

20. $\dfrac{5}{10} = \dfrac{n}{100}$

21. $\dfrac{15}{n} = \dfrac{3}{24}$

22. $\dfrac{4}{8} = \dfrac{1}{n}$

23. A compact car rents for 3 days for $132. How much does it cost to rent the car for 1 day?

24. A mid-size car rents for 5 days for $300. How much does it cost to rent the car for 2 days?

Answer _____

Answer _____

Writing Percents

Percent means hundredths. When using percents, the whole is divided into 100 equal parts. Twenty-five percent (25%) means 25 hundredths or 25 out of 100 parts. The sign % is read *percent*.

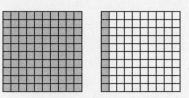

25%
twenty-five percent

100%
one hundred percent

110%
one hundred ten percent

Use These Steps

Write ten percent using the percent sign.

1. Write ten.

10

2. Write the percent sign after the 10.

10%

Write each percent using the percent sign.

1. five percent = 5%

2. eight percent =

3. two percent =

4. seven percent =

5. sixteen percent =

6. twenty-four percent =

7. thirty-nine percent =

8. forty-one percent =

9. sixty percent =

10. fifty percent =

11. eighty percent =

12. one hundred percent =

13. four and one half percent = $4\frac{1}{2}\%$

14. thirty-three and one third percent =

15. two hundred forty-five percent =

16. thirteen percent =

17. two and three tenths percent =

18. twenty and one half percent =

Write a percent for each figure.

19.

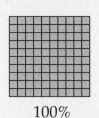

2%

20.

21.

22.

Changing Percents to Decimals

You can write any percent as a decimal by writing the number without the percent sign and moving the decimal point two places to the left. Remember that you can drop a zero at the end of a decimal without changing its value.

$$50\% = 50 = .5 \qquad 36\% = 36 = .36 \qquad 12\% = 12 = .12$$

For percents less than 10%, put a zero in front of the number so that you can move the decimal point two places to the left.

$$1\% = 01 = .01 \qquad 5\% = 05 = .05 \qquad 9\% = 09 = .09$$

Use These Steps

Change 75% to a decimal.

1. Write the number without the percent sign.

   75

2. Move the decimal point two places to the left.

   $75 = .75$

Change each percent to a decimal.

1. $35\% = 35 = .35$

2. $19\% =$

3. $27\% =$

4. $42\% =$

5. $10\% = 10 = .10 = .1$

6. $30\% =$

7. $70\% =$

8. $90\% =$

9. $6\% = 06 = .06$

10. $1\% =$

11. $3\% =$

12. $9\% =$

13. $29\% =$

14. $4\% =$

15. $40\% =$

16. $72\% =$

17. $54\% =$

18. $2\% =$

19. $41\% =$

20. $66\% =$

21. Marco spends 25% of his work day working for a messenger service. Write as a decimal the part of each work day he spends working for the messenger service.

22. The other 75% of his work day, Marco plays drums for a band. Write as a decimal the part of the day he plays in the band.

Answer_____

Answer_____

Changing Percents Greater Than 100% to Decimals

If you have 100% of something, you have the whole thing. Sometimes, however, you will have percents greater than 100%. To change percents greater than 100% to a decimal, write the number without the percent sign and move the decimal two places to the left.

$$200\% = 200 = 2.00 = 2 \qquad 250\% = 250 = 2.50 = 2.5 \qquad 309\% = 309 = 3.09$$

Use These Steps

Change 120% to a decimal.

1. Write the number without the percent sign.

2. Move the decimal point two places to the left. Drop the zero at the end.

120

$$1\,20 = 1.20 = 1.2$$

Change each percent to a decimal.

1.
$$300\% = 300 = 3.00 = 3$$

2.
$$500\% =$$

3.
$$700\% =$$

4.
$$270\% = 270 = 2.70 = 2.7$$

5.
$$110\% =$$

6.
$$450\% =$$

7.
$$107\% = 107 = 1.07$$

8.
$$308\% =$$

9.
$$701\% =$$

10.
$$925\% =$$

11.
$$714\,\% =$$

12.
$$283\% =$$

13.
$$520\% =$$

14.
$$624\% =$$

15.
$$800\% =$$

16.
$$135\% =$$

17.
$$470\% =$$

18.
$$209\% =$$

19. There was a 212% increase in the number of houses sold last year. Write 212% as a decimal.

20. A construction company built 145% more new houses this year than last year. Write 145% as a decimal.

Answer _____

Answer _____

Changing Complex Percents to Decimals

A complex percent has a whole number and a fraction. To change a complex percent to a decimal, change the fraction to a decimal. Then move the decimal point two places to the left. Add a zero if necessary.

$$6\tfrac{1}{2}\% = 6.5\% = 06.5 = .065$$

Some complex percents do not have exact decimal equivalents. When you divide to change the percent to a decimal, you will always have a remainder. Write the remainder as a fraction.

$$33\tfrac{1}{3}\% = 33\tfrac{1}{3} = .33\tfrac{1}{3}$$

Use These Steps

Change $10\tfrac{1}{2}\%$ to a decimal.

1. Change the fraction to a decimal.

    $$10\tfrac{1}{2}\% = 10.5\%$$

2. Write the number without the percent sign.

    $$10.5$$

3. Move the decimal point two places to the left.

    $$10.5 = .105$$

Change each percent to a decimal.

1.
$$2\tfrac{1}{4}\% = 2.25\% = 02.25 = .0225$$

2.
$$3\tfrac{1}{10}\% =$$

3.
$$6\tfrac{1}{2}\% =$$

4.
$$7\tfrac{1}{3}\% = 07\tfrac{1}{3} = .07\tfrac{1}{3}$$

5.
$$10\tfrac{2}{3}\% =$$

6.
$$2\tfrac{1}{5}\% =$$

7.
$$30\tfrac{1}{2}\% =$$

8.
$$16\tfrac{3}{10}\% =$$

9.
$$50\tfrac{1}{5}\% =$$

10.
$$8\tfrac{1}{2}\% =$$

11.
$$22\tfrac{1}{10}\% =$$

12.
$$1\tfrac{21}{25}\% =$$

13.
$$9\tfrac{9}{10}\% =$$

14.
$$2\tfrac{4}{5}\% =$$

15.
$$25\tfrac{4}{25}\% =$$

When you borrow money from a bank or credit union, you must pay interest on the money you borrow. When you save money, the bank or credit union pays you interest on the money you save. Sometimes the bank or credit union pays you interest on the money in your checking account. Interest is figured using a rate or percent.

Example The interest rate for a four year new car loan is $9\frac{1}{4}\%$ at National Bank. The interest rate at Western Bank is 9.5%. Which bank has the lower interest rate?

$9\frac{1}{4}\% = 09.25 = .0925$

$9.5\% = 09.5 = .095$

$.0925 < .095$, so $9\frac{1}{4}\% < 9.5\%$.

The interest rate at National Bank is lower.

Solve.

1. The interest rate for a used-car loan for four years at National Bank is $10\frac{1}{2}\%$. The interest rate at Western Bank is 10.25%. Which bank has the lower interest rate?

   Answer_____

2. The interest rate for a used-car loan for three years at National Bank is $10\frac{3}{4}\%$. The interest rate at Western Bank is 10.5%. Which bank has the lower interest rate?

   Answer_____

3. The interest rate for a thirty-year conventional home loan at National Bank is $8\frac{1}{4}\%$. The interest rate at Western Bank is 8.75%. Which bank has the lower interest rate?

   Answer_____

4. The interest rate for a thirty-year FHA home loan at National Bank is $8\frac{1}{4}\%$. The interest rate at Western Bank is 8.125%. Which bank has the lower interest rate?

   Answer_____

Changing Decimals to Percents

To change a decimal to a percent, move the decimal point two places to the right. This is the same as multiplying by 100. You may need to add a zero to the end of the number. Add a percent sign.

$$.06 = .06 = 6\% \qquad .5 = .50 = 50\% \qquad .66\frac{2}{3} = .66\frac{2}{3} = 66\frac{2}{3}\%$$

Use These Steps

Change .2 to a percent.

1. Add a zero.

2. Move the decimal point two places to the right.

3. Add a percent sign.

$$.2 = .20 \qquad\qquad .20 = .20 = 20 \qquad\qquad 20\%$$

Change each decimal to a percent.

1. $.01 = .01 = 1\%$

2. $.07 =$

3. $.09 =$

4. $.05 =$

5. $.1 = .10 = 10\%$

6. $.3 =$

7. $.6 =$

8. $.9 =$

9. $.25 =$

10. $.75 =$

11. $.47 =$

12. $.65 =$

13. $.865 =$

14. $.031 =$

15. $.799 =$

16. $.209 =$

17. $.33\frac{1}{3} = .33\frac{1}{3} = 33\frac{1}{3}\%$

18. $.16\frac{2}{3} =$

19. $.83\frac{1}{3} =$

20. $.16\frac{2}{3} =$

21. $.99 =$

22. $.08 =$

23. $.074 =$

24. $.5 =$

25. $.033 =$

26. $.7 =$

27. $.06 =$

28. $.225 =$

Real-Life Application **Daily Living**

Many real-life situations use percents. Taxes are figured using percents. Some sale prices are shown as a "percent off." Interest on loans and savings accounts is a percent of the amount borrowed or saved.

 Example You and your employer contribute a total of 15% of your wages for social security. Write this amount as a decimal.

$$15\% = 15 = .15$$

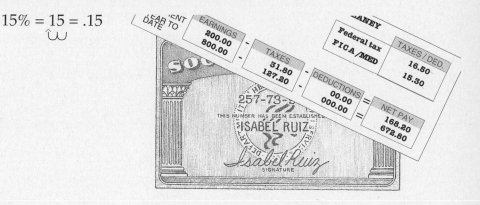

Write each percent as a decimal.

1. Some people pay 28% of their taxable income for income taxes.

   Answer_____

2. Sales tax in one large city is 8%.

   Answer_____

3. Gasoline tax in one state is 22%.

   Answer_____

4. At last week's sale, Janie saved 25%.

   Answer_____

5. Barbara earns 5% interest on her savings account.

   Answer_____

6. Sam pays 16% interest on his credit card.

   Answer_____

7. Oscar pays 9% interest on his student loan.

   Answer_____

8. Julia makes 3% commission on each car she sells.

   Answer_____

9. Jan saves 10% of her pay each month.

   Answer_____

10. Lyla got a 4% raise this year.

    Answer_____

Mixed Review

Write a ratio for each problem. Reduce if possible.

1. 3 tables to 12 chairs

2. 9 buses for 180 children

3. 5 pounds for $3

4. 4 tires for 1 car

Solve each proportion for n.

5. $\frac{3}{7} = \frac{n}{14}$

6. $\frac{2}{n} = \frac{1}{9}$

7. $\frac{4}{5} = \frac{12}{n}$

8. $\frac{n}{25} = \frac{12}{5}$

Write each percent with the percent sign.

9. twenty-five percent =

10. one hundred percent =

11. two and one half percent =

12. thirty percent =

Change each percent to a decimal.

13. 29% =

14. 8% =

15. $2\frac{1}{2}\% =$

16. $1\frac{9}{10}\% =$

17. 7.2% =

18. 75% =

19. 120% =

20. 406% =

21. $5\frac{1}{2}\% =$

22. $8\frac{1}{3}\% =$

23. $10\frac{2}{3}\% =$

24. 19.5% =

Change each decimal to a percent.

25. .8 =

26. .05 =

27. .16 =

28. .425 =

29. .1 =

30. .75 =

31. .081 =

32. .01 =

33. .506 =

34. .22 =

35. .6 =

36. .947 =

Real-Life Application

Many statistics in sports are given in percents. The batting averages in baseball are really percents. The chart to the right shows the batting averages of several baseball players. Batting averages are written with three decimal places.

Batting Averages	Player	Hits	Times at Bat	Average	Percent
	Ruiz, J	35	125	.280	28%
	Lee, A.	4	20		
	Delsanto, S.	18	75		
	Montoya, B.	3	12		
	Shaffer, T.	7	25		

Example Ruiz has a ratio of 35 hits out of 125 times at bat. What is Ruiz's batting average? What is the percent?

$$\frac{\text{hits}}{\text{times at bat}} = \frac{35}{125}$$

$$\frac{35}{125} = 35 \div 125$$

$$35 \div 125 = .28$$

$$.28 = .28 = 28\%$$

Ruiz has a batting average of .280 and a percent of 28%.

Solve. Fill in the chart.

1. What is Lee's batting average? What is the percent?

   Answer_____

2. What is Delsanto's batting average? What is the percent?

   Answer_____

3. What is Montoya's batting average? What is the percent?

4. What is Shaffer's batting average? What is the percent?

Changing Percents to Fractions

For some percent problems, it may be easier to change the percent to a fraction instead of to a decimal.

Since percent means hundredths, to change a percent to a fraction, write the number over 100 without the percent sign. When you have a percent greater than 100%, you will get an improper fraction.

$$25\% = \frac{25}{100} = \frac{1}{4} \qquad 37\% = \frac{37}{100} \qquad 250\% = \frac{250}{100} = \frac{5}{2}$$

Use These Steps

Change 50% to a fraction.

1. Write the number over 100 without the percent sign.

$$50\% = \frac{50}{100}$$

2. Reduce.

$$\frac{5\cancel{0}}{10\cancel{0}} = \frac{5}{10} = \frac{1}{2}$$

Change each percent to a fraction. Reduce if possible.

1. $75\% = \frac{75}{100} = \frac{3}{4}$

2. $16\% =$

3. $45\% =$

4. $99\% =$

5. $30\% =$

6. $50\% =$

7. $6\% =$

8. $2\% =$

9. $5\% =$

10. $25\% =$

11. $40\% =$

12. $83\% =$

Change each percent to an improper fraction. Reduce if possible.

13. $300\% = \frac{300}{100} = \frac{3}{1}$

14. $500\% =$

15. $870\% =$

16. $225\% =$

17. $460\% =$

18. $986\% =$

19. $200\% =$

20. $670\% =$

21. Koji lost 60% of his customers when they switched to another telephone company. Express as a fraction the percent of customers lost.

22. Koji increased his sales by 120% for the year by convincing most of his remaining customers to buy additional services. Express as a fraction the percent of increased sales.

Answer_____

Answer_____

Changing Complex Percents to Fractions

When you change a complex percent to a fraction, write the mixed number over 100 without the percent sign. Divide the numerator by the denominator.

Use These Steps

Change $3\frac{1}{3}\%$ to a fraction.

1. Write the mixed number over 100 without the percent sign. Set up a division problem.

$$3\frac{1}{3}\% = \frac{3\frac{1}{3}}{100} = 3\frac{1}{3} \div 100$$

2. Change both numbers to improper fractions. Invert and multiply. Reduce.

$$3\frac{1}{3} \div 100 = \frac{10}{3} \div \frac{100}{1} = \frac{\overset{1}{\cancel{10}}}{3} \times \frac{1}{\underset{10}{\cancel{100}}} = \frac{1}{30}$$

Change each percent to a fraction.

1.

$$7\frac{1}{2}\% =$$

$$\frac{7\frac{1}{2}}{100} = 7\frac{1}{2} \div 100 = \frac{15}{2} \div \frac{100}{1} = \frac{\overset{3}{\cancel{15}}}{2} \times \frac{1}{\underset{20}{\cancel{100}}} = \frac{3}{40}$$

2.

$$12\frac{1}{2}\% =$$

3.

$$8\frac{1}{3}\% =$$

4.

$$66\frac{2}{3}\% =$$

5.

$$16\frac{2}{3}\% =$$

6.

$$33\frac{1}{3}\% =$$

7.

$$1\frac{1}{4}\% =$$

8.

$$3\frac{3}{4}\% =$$

9.

$$3\frac{1}{8}\% =$$

10.

$$9\frac{3}{8}\% =$$

Changing Fractions to Percents

To change a fraction to a percent, first change the fraction to a decimal by dividing the numerator by the denominator. Move the decimal point two places to the right and add a percent sign.

Use These Steps

Change $\frac{3}{4}$ to a percent.

1. Change the fraction to a decimal by dividing the numerator by the denominator. Divide until there is no remainder.

$$\begin{array}{r} .75 \\ 4\overline{)3.00} \\ -28 \\ \hline 20 \\ -20 \\ \hline 0 \end{array}$$

2. Move the decimal point two places to the right. Add a percent sign.

$$.75 = .75 = 75\%$$

Change each fraction to a percent.

1. $\frac{1}{2} =$

$$\begin{array}{r} .5 \\ 2\overline{)1.0} \\ -10 \\ \hline 0 \end{array}$$

$.5 = .50 = 50\%$

2. $\frac{3}{15} =$

3. $\frac{2}{5} =$

4. $\frac{7}{10} =$

5. $\frac{9}{25} =$

6. $\frac{7}{20} =$

7. $\frac{3}{20} =$

8. $\frac{7}{25} =$

9. $\frac{1}{5} =$

10. $\frac{16}{25} =$

11. $\frac{1}{4} =$

12. $\frac{3}{10} =$

13. $\frac{9}{10} =$

14. $\frac{3}{5} =$

15. $\frac{9}{150} =$

16. $\frac{3}{100} =$

Changing Fractions to Percents

When changed to decimals, some fractions will always have a remainder. When this happens, keep only two places in the answer and write the remainder as a fraction. Then move the decimal point two places to the right and add a percent sign.

Use These Steps

Change $\frac{1}{3}$ to a percent.

1. Change the fraction to a decimal with two places by dividing the numerator by the denominator. Write the remainder as a fraction.

$$\begin{array}{r} .33\frac{1}{3} \\ 3\overline{)1.00} \\ -\ 9 \\ \hline 10 \\ -\ 9 \\ \hline 1 \end{array}$$

2. Move the decimal point two places to the right. Add a percent sign.

$$.33\frac{1}{3} = .33\frac{1}{3} = 33\frac{1}{3}\%$$

Change each fraction to a percent. Reduce if possible.

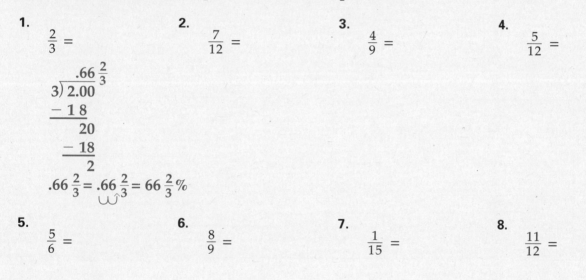

1. $\frac{2}{3} =$

$$\begin{array}{r} .66\frac{2}{3} \\ 3\overline{)2.00} \\ -\ 1\ 8 \\ \hline 20 \\ -\ 18 \\ \hline 2 \end{array}$$

$$.66\frac{2}{3} = .66\frac{2}{3} = 66\frac{2}{3}\%$$

2. $\frac{7}{12} =$

3. $\frac{4}{9} =$

4. $\frac{5}{12} =$

5. $\frac{5}{6} =$

6. $\frac{8}{9} =$

7. $\frac{1}{15} =$

8. $\frac{11}{12} =$

Fractions, Decimals, and Percents

You will find it helpful to memorize these commonly used equivalent fractions, decimals, and percents. This will save you some work when doing problems with percents.

Fill in the chart below.

Fraction	Decimal	Percent
$\frac{1}{100}$	.01	1%
$\frac{1}{10}$		
$\frac{1}{8}$	.125	$12\frac{1}{2}\%$
$\frac{1}{6}$	$.16\frac{2}{3}$	$16\frac{2}{3}\%$
$\frac{1}{5}$		
$\frac{1}{4}$		
$\frac{3}{10}$		
$\frac{1}{3}$		
$\frac{3}{8}$		
$\frac{2}{5}$		
$\frac{1}{2}$		
$\frac{3}{5}$		
$\frac{5}{8}$		
$\frac{2}{3}$		
$\frac{7}{10}$		
$\frac{3}{4}$		
$\frac{5}{6}$		
$\frac{7}{8}$		
$\frac{9}{10}$		
1	1.0	100%

Comparing Decimals and Percents

If you want to compare a decimal and a percent, they must be in the same form.

Remember the symbol < means *less than*.
                     > means *greater than*.

Use These Steps

Compare .2 ☐ 25%

1. You can change .2 to a percent or you can change 25% to a decimal.

$$.2 = .20 = 20\%$$
or
$$25\% = 25 = .25$$

2. Compare the new forms.

$$20\% < 25\%, \text{ so } .2 < 25\%$$
or
$$.2 < .25, \text{ so } .2 < 25\%$$

Change each pair of numbers to the same form. Then compare. Use <, >, or =.

1. 14% $\boxed{<}$ 2.1
   14% = 14 = .14
   .14 < 2.1, so 14% < 2.1

2. 104% $\boxed{<}$ 1.4
   1.4 = 1.40 = 140%
   104% < 140%, so 104% < 1.4

3. 8% ☐ .026

4. 205% ☐ .25

5. 36% ☐ .36

6. 70% ☐ .9

7. .06 ☐ 2.6%

8. 1.37 ☐ 14.2%

9. .875 ☐ 75%

10. 1.3 ☐ 158%

11. 6.2 ☐ $62\frac{1}{2}\%$

12. .013 ☐ 13%

13. .125 ☐ $12\frac{1}{2}\%$

14. .31 ☐ 31%

15. 12.3 ☐ 1.5%

16. .6 ☐ 60%

Comparing Fractions and Percents

Fractions and percents must be in the same form if you want to compare them. The easiest way to compare fractions and percents is to change each number to a decimal.

Use These Steps

Compare $\frac{1}{2}$ ☐ 5%.

1. Change each number to a decimal.

$$\frac{1}{2} = .5$$

$$5\% = .005$$

2. Compare the decimals.

$$.5 > .005, \text{ so } \frac{1}{2} > 5\%$$

Compare. You may want to use the chart on page 94 to help you. Write <, >, or =.

1.
$$\frac{2}{5} \boxed{>} 20\%$$
$$\frac{2}{5} = .4$$
$$20\% = .20$$
$$.4 > .20, \text{ so } \frac{2}{5} > 20\%$$

2.
$$\frac{1}{8} \ \square \ 12\%$$

3.
$$\frac{1}{3} \ \square \ 33\frac{1}{3}\%$$

4.
$$10\% \ \square \ \frac{1}{4}$$

5.
$$95\% \ \square \ \frac{3}{4}$$

6.
$$1\frac{1}{2}\% \ \square \ \frac{1}{2}$$

7.
$$\frac{1}{5} \ \square \ 30\%$$

8.
$$41\% \ \square \ \frac{7}{10}$$

9.
$$\frac{5}{6} \ \square \ 83\frac{1}{3}\%$$

10.
$$25\% \ \square \ \frac{1}{4}$$

11.
$$\frac{3}{5} \ \square \ 50\%$$

12.
$$75\% \ \square \ \frac{2}{3}$$

13.
$$\frac{1}{6} \ \square \ 20\%$$

14.
$$33\frac{1}{3}\% \ \square \ \frac{3}{8}$$

15.
$$\frac{5}{6} \ \square \ 85\%$$

16.
$$\frac{5}{8} \ \square \ 60\%$$

17. Kai can buy a couch for $\frac{1}{4}$ off or 20% off. Which is the better offer: $\frac{1}{4}$ off or 20% off?

18. The price of a chair Kai likes is marked down 25% in one store, and $\frac{1}{3}$ off in another store. Which is the better value: 25% off or $\frac{1}{3}$ off?

Answer_____

Answer_____

Problem Solving: Using a Budget

This budget shows the average amount that Creative Cleaners spends each month on equipment, taxes, advertising, telephone expenses, transportation, insurance, and miscellaneous expenses.

$ 2,400	Equipment
3,720	Taxes
600	Miscellaneous
1,200	Advertising
1,440	Telephone
1,800	Transportation
840	Insurance
$12,000	

If you change each amount to a percent, you will find out what part of Creative Cleaners' budget is spent on each item.

Example Creative Cleaners' budget this month was $12,000. They spent $2,400 on equipment. What percent of their budget did they spend on equipment?

▶ **Step 1.** Write the amount spent for equipment over the amount of the total budget.

$$\frac{2,400}{12,000}$$

▶ **Step 2.** Reduce to lowest terms and change the fraction to a percent.

$$\frac{24\cancel{00}}{120\cancel{00}} = \frac{24}{120} = \frac{1}{5}$$

$$\frac{1}{5} = .20 = 20\%$$

Creative Cleaners spent 20% of their budget on equipment.

Write a percent for each problem.

1. Creative Cleaners spent $3,720 on taxes. What percent of the budget did they spend on taxes?

2. Creative Cleaners spent $600 on miscellaneous expenses. What percent of the budget did they spend on miscellaneous expenses?

Answer_____

Answer_____

3. Creative Cleaners spent $1,200 on advertising. What percent of the budget did they spend on advertising?

Answer_____

4. Creative Cleaners spent $1,440 on telephone expenses. What percent of the budget did they spend on telephone expenses?

Answer_____

5. What percent of the budget did Creative Cleaners spend on transportation?

Answer_____

6. What percent of the budget did Creative Cleaners spend on insurance?

Answer_____

7. About $720 of the money for telephone expenses was spent to buy a cellular phone. What percent of their budget is this?

Answer_____

8. Creative Cleaners spent $720 for a cellular phone. What percent of their money for telephone expenses is this?

Answer_____

9. Creative Cleaners spent $360 on newspaper ads. What percent of their advertising money is this?

Answer_____

10. Creative Cleaners spent $360 on newspaper ads. What percent of their budget is this?

Answer_____

11. Creative Cleaners spent $1,080 of their transportation expenses on gasoline. What percent of their transportation expenses is this?

Answer_____

12. Creative Cleaners spent $1,080 on gasoline. What percent of their budget is this?

Answer_____

Unit 4 *Review*

Write a ratio for each problem. Reduce if possible.

1. 4 bedrooms to 3 bathrooms

2. 5 hotdogs and 10 hamburgers

3. 3 chances out of 100

4. 2 quarts of oil for $3

Solve each proportion.

5. $\frac{1}{2} = \frac{n}{6}$

6. $\frac{2}{8} = \frac{1}{n}$

7. $\frac{n}{70} = \frac{1}{10}$

8. $\frac{6}{n} = \frac{3}{4}$

Write each percent using a percent sign.

9. three percent =

10. ninety-nine percent =

11. six and one third percent =

12. one hundred two percent =

Change each percent to a decimal.

13. 45% =

14. 80% =

15. 2% =

16. $33\frac{1}{3}\% =$

17. 240% =

18. $9\frac{1}{2}\% =$

19. .8% =

20. $66\frac{2}{3}\% =$

Change each decimal to a percent.

21. .3 =

22. .09 =

23. .117 =

24. .25 =

25. 1.8 =

26. 1.01 =

27. .6 =

28. .99 =

Change each percent to a fraction.

29. 30% =

30. 9% =

31. 28% =

32. 300% =

33. 75% =

34. $8\frac{1}{3}\% =$

35. $66\frac{2}{3}\% =$

36. 120% =

Change each fraction to a percent.

37.
$$\frac{1}{4} =$$

38.
$$\frac{2}{3} =$$

39.
$$1\frac{1}{2} =$$

40.
$$6\frac{7}{20} =$$

41.
$$\frac{3}{10} =$$

42.
$$\frac{4}{25} =$$

43.
$$7\frac{3}{4} =$$

44.
$$9\frac{1}{3} =$$

Change each fraction to a decimal and a percent.

45.
$$\frac{1}{100} =$$

46.
$$\frac{1}{2} =$$

47.
$$\frac{2}{3} =$$

48.
$$\frac{1}{4} =$$

49.
$$\frac{3}{4} =$$

50.
$$\frac{1}{3} =$$

51.
$$\frac{7}{10} =$$

52.
$$\frac{1}{8} =$$

Compare. Use <, >, or =.

53.
$$10\% \ \square \ \frac{1}{10}$$

54.
$$.5 \ \square \ 5\%$$

55.
$$\frac{3}{4} \ \square \ 40\%$$

56.
$$26\% \ \square \ 3.1$$

Below is a list of the problems in this review and the pages on which the skills are taught. If you missed any problems, turn to the pages listed and practice the skills. Then correct the problems you missed in the Unit Review.

Problems	Pages		Problems	Pages
1-4	75		29-36	91
5-8	76-78		37-44	92-93
9-12	81		45-52	94
13-20	82-84		53-56	95-96
21-28	86			

Unit 5 USING PERCENTS

We use percents all the time. We use them to compute interest on loans and savings accounts, income taxes, sales tax, tips, sales commissions, and pay raises.

In this unit, you will learn how to solve percent problems and how to use percents in some familiar real-life situations.

Getting Ready

You should be familiar with the skills on this page and the next before you begin this unit. To check your answers, turn to page 186.

You may need to change from a percent to a decimal before you work a problem. Remember, percent means hundredths.

Change each percent to a decimal.

1.
$15\% = 15. = .15$

2.
$9\frac{1}{4}\% =$

3.
$7\% =$

4.
$5\% =$

5.
$20\% =$

6.
$33\% =$

7.
$8.3\% =$

8.
$99\% =$

9.
$105\% =$

10.
$65\% =$

11.
$3\frac{1}{2}\% =$

12.
$14.6\% =$

13.
$300\% =$

14.
$16\% =$

15.
$10\% =$

16.
$250\% =$

For review, see Unit 4, pages 82–84.

Getting Ready

▶ Sometimes you'll have to change a percent to a fraction.

Change each percent to a fraction. Reduce if possible.

17.
$10\% = \frac{10}{100} = \frac{1}{10}$

18.
$1\% =$

19.
$20\% =$

20.
$25\% =$

21.
$33\frac{1}{3}\% =$

22.
$66\frac{2}{3}\% =$

23.
$50\% =$

24.
$60\% =$

25.
$40\% =$

26.
$75\% =$

27.
$80\% =$

28.
$100\% =$

For review, see Unit 4, pages 90-91.

▶ The answers to some percent problems may need to be rounded.

Round each decimal to the nearest hundredth.

29.
$1.072 = 1.07$

30.
$.333 =$

31.
$4.625 =$

32.
$10.001 =$

33.
$4.366 =$

34.
$.695 =$

35.
$12.361 =$

36.
$186.001 =$

37.
$20.455 =$

38.
$9.929 =$

39.
$340.016 =$

40.
$829.405 =$

For review, see Unit 1, page 27.

▶ Often when you work percent problems, you will need to multiply or divide with decimals.

Solve.

41.
$1.23 \times .15 =$

$$\begin{array}{r} 1.23 \\ \times \ .15 \\ \hline 6\ 15 \\ +\ 12\ 3 \\ \hline .18\ 45 \end{array}$$

42.
$.05 \times 72 =$

43.
$84 \times .25 =$

44.
$30 \times .1 =$

45.
$10 \div .5 =$

46.
$84 \div .21 =$

47.
$15.4 \div 20 =$

48.
$3.6 \div .4 =$

For review, see Unit 3, pages 52-56, 62-68.

Understanding Percent Problems

There are three pieces to a percent problem: *the part*, *the whole*, and *the percent*.

10% of 250 is 25

part = 25
whole = the total, 250
percent = the number with the % sign, 10%

Use These Steps

Find the part, the whole, and the percent in: 50% of 150 is 75.

1. Write the part.

part = 75

2. Write the whole.

whole = 150

3. Write the percent.

percent = 50%

Find the part, the whole, and the percent.

1.
40% of 200 is 80
part = 80
whole = 200
percent = 40%

2.
30% of 60 is 18

3.
90% of 50 is 45

4.
100% of 30 is 30

5.
5% of 500 is 25

6.
8% of 1,200 is 96

7.
150% of 100 is 150

8.
200% of 50 is 100

9.
500% of 20 is 100

10.
12.5% of 80 is 10

11.
$33\frac{1}{3}$% of 60 is 20

12.
4.3% of 1,000 is 43

13. 15% off the original price of a $12,000 car is $1,800.

14. A loan for 80% of the $16,000 sticker-price of a truck is $12,800.

Answer_____

Answer_____

Understanding Percent Problems: The Part

Sometimes in a percent problem, one of the three pieces is missing. You can find the missing piece by using the percent triangle. The percent triangle shows how the three pieces are related.

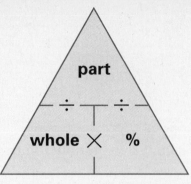

The Percent Triangle

To solve for the part, cover the word *part*. The remaining pieces are connected by a multiplication sign. Multiply the pieces you have to find the part.

$$part = whole \times percent$$

Use These Steps

Find the part, the whole, and the percent in: 10% of 250 is what?

1. Write the pieces you have.

whole = 250
percent = 10%

2. To find the part, write a percent sentence.

part = whole × percent

3. Replace the words with the numbers for each piece.

part = 250 × 10%

Find the part, the whole, and the percent.

1.
65% of 40 is what?
whole = 40
percent = 65%
part = 40 × 65%

2.
32% of 19 is what?

3.
6% of 10 is what?

4.
1% of 327 is what?

5.
250% of 700 is what?

6.
$1\frac{1}{2}$% of 36 is what?

7.
What is 50% of 80?
whole = 80
percent = 50%
part = 80 × 50%

8.
What is $33\frac{1}{3}$% of 9?

9.
What is 125% of 86?

10.
What is 10% of 65?

11.
What is $33\frac{1}{3}$% of 500?

12.
What is 75% of 350?

Understanding Percent Problems: The Whole

In a percent problem, the missing piece may be the whole. Use the percent triangle the same way you did to find the part.

To solve for the whole, cover the word *whole*. The remaining pieces are connected by a division sign. Divide the pieces you have to find the whole.

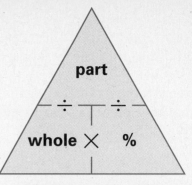

The Percent Triangle

$$\text{whole} = \text{part} \div \text{percent}$$

Use These Steps

Find the part, the whole, and the percent in: **24 is 30% of what number?**

1. Write the pieces you have.

   part = 24
   percent = 30%

2. To find the whole, write a percent sentence.

   whole = part ÷ percent

3. Replace the words with the numbers for each piece.

   whole = 24 ÷ 30%

Find the part, the whole, and the percent.

1.
3 is 10% of what number?
part = 3
percent = 10%
whole = 3 ÷ 10%

2.
10 is 25% of what number?

3.
24 is 3% of what number?

4.
15 is 5% of what number?

5.
33 is 110% of what number?

6.
40 is $66\frac{2}{3}$% of what number?

7.
20 is 25% of what number?

8.
7 is 10% of what number?

9.
400 is 200% of what number?

10.
75 is $33\frac{1}{3}$% of what number?

11.
90 is 25% of what number?

12.
130 is 160% of what number?

Understanding Percent Problems: The Percent

In a percent problem, the missing piece may be the percent. Use the percent triangle the same way you did to find the part and the whole.

To solve for the percent, cover the percent sign. The remaining pieces are connected by a division sign. Divide the pieces you have to find the percent.

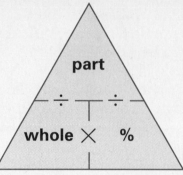

$$\text{percent} = \text{part} \div \text{whole}$$

The Percent Triangle

Use These Steps

Find the part, the whole, and the percent in: What percent of 40 is 20?

1. Write the pieces you have.

   part = 20
   whole = 40

2. To find the percent, write a percent sentence.

   percent = part ÷ whole

3. Replace the words with the numbers for each piece.

   percent = 20 ÷ 40

Find the part, the whole, and the percent.

1.
   What percent of 16 is 4?
   part = 4
   whole = 16
   percent = 4 ÷ 16

2.
   What percent of 20 is 10?

3.
   What percent of 120 is 4?

4.
   What percent of 100 is 1?

5.
   What percent of 36 is 72?

6.
   What percent of 4 is 8?

7.
   10 is what percent of 12?
   part = 10
   whole = 12
   percent = 10 ÷ 12

8.
   3 is what percent of 2?

9.
   75 is what percent of 25?

10.
    12 is what percent of 3?

11.
    150 is what percent of 200?

12.
    15 is what percent of 25?

Finding the Part: Changing Percents to Decimals

Remember, you can find the part in a percent problem by using this sentence:

$$\text{part} = \text{whole} \times \text{percent}$$

Change the percent to a decimal. Then multiply to find the part.

Use These Steps

What is 15% of 900?

1. Write the pieces.

   whole = 900
   percent = 15%
   part = 900 × 15%

2. Change the percent to a decimal.

   15% = 15. = .15

3. Multiply 900 by .15 to find the part.

   900 × .15 = 135

Find the part.

1.
   What is 10% of 20?
   whole = 20
   percent = 10% = .1
   part = 20 × .1 = 2

2. What is 95% of 300?

3. What is 5% of 60?

4. What is 22% of 50?

5. What is 36% of 125?

6. What is 60% of 250?

7.
   15% of 280 is what?
   whole = 280
   percent = 15% = .15
   part = 280 × .15 = 42

8. 12% of 300 is what?

9. 5% of 80 is what?

10. Richard pays 15% for federal taxes. He earns $1,000 each month. How much federal tax does he pay each month?

11. Richard pays a state tax of 4% on the money he earns. How much did he pay for state taxes?

Answer_____

Answer_____

Finding the Part: Changing Complex Percents to Decimals

In some problems, the percent you're working with may be a complex percent. To solve the problem, change the percent to a decimal.

Use These Steps

What is $1\frac{1}{2}$% of 400?

1. Write the pieces.

    whole = 400
    percent = $1\frac{1}{2}$%
    part = $400 \times 1\frac{1}{2}$%

2. Change the percent to a decimal.

    $1\frac{1}{2}$% = 1.5% = 01.5 = .015

3. Multiply 400 by .015 to find the part.

    $400 \times .015 = 6$

Find the part.

1.
   What is $8\frac{1}{5}$% of 500?
   whole = 500
   percent = $8\frac{1}{5}$% = .082
   part = $500 \times .082 = 41$

2.
   What is $16\frac{1}{2}$% of 400?

3.
   What is $24\frac{1}{4}$% of 400?

4.
   What is $60\frac{3}{4}$% of 800?

5.
   What is $4\frac{2}{5}$% of 250?

6.
   What is $7\frac{3}{10}$% of 1,000?

7.
   $62\frac{1}{2}$% of 480 is what?
   whole = 480
   percent = $62\frac{1}{2}$% = .625
   part = $480 \times .625 = 300$

8.
   $7\frac{3}{5}$% of 500 is what?

9.
   $12\frac{1}{2}$% of 24 is what?

10.
   $9\frac{4}{5}$% of 1,500 is what?

Finding the Part: Changing Percents to Fractions

When solving for the part, it may be easier to change the percent to a fraction instead of to a decimal. Remember, a percent can be changed to a fraction by dropping the percent sign and writing the percent as the numerator of a fraction with 100 as the denominator.

The table on page 94 lists some common percents and fractions.

Use These Steps

Find the three pieces in: What is 50% of 24?

1. Write the pieces.

2. Change the percent to a fraction. Reduce.

3. Multiply 24 by $\frac{1}{2}$ to find the part.

whole = 24
percent = 50%
part = 24 × 50%

$50\% = \frac{50}{100} = \frac{1}{2}$

$\frac{\overset{12}{\cancel{24}}}{1} \times \frac{1}{\underset{1}{\cancel{2}}} = \frac{12}{1} = 12$

Find the part.

1.
What is 30% of 20?
whole = 20
percent = 30% $= \frac{3\cancel{0}}{10\cancel{0}} = \frac{3}{10}$
part $= \frac{\overset{2}{\cancel{20}}}{1} \times \frac{3}{\underset{1}{\cancel{10}}} = \frac{6}{1} = 6$

2.
What is 25% of 32?

3.
What is 15% of 240?

4.
10% of 500 is what?
whole = 500
percent = 10% $= \frac{1\cancel{0}}{10\cancel{0}} = \frac{1}{10}$
part $= \frac{\overset{50}{\cancel{500}}}{1} \times \frac{1}{\underset{1}{\cancel{10}}} = \frac{50}{1} = 50$

5.
50% of 80 is what?

6.
20% of 125 is what?

7. Sybil and Dan spent $250 one month on home repairs. They spent 24% of the amount replacing the front step. How much money did they spend on the steps?

8. Sybil and Dan spent 60% of the $250 fixing the roof. How much money did they spend on the roof?

Answer_____

Answer_____

Finding the Part: Changing Complex Percents to Fractions

When you solve problems that use percents such as $33\frac{1}{3}\%$ and $66\frac{2}{3}\%$ that don't have an exact decimal equivalent, use their fraction equivalent. You may want to use the table on page 94 to find some of the common fraction equivalents.

Use These Steps

What is $33\frac{1}{3}\%$ of 24?

1. Write the pieces.

   whole = 24
   percent = $33\frac{1}{3}\%$
   part = $24 \times 33\frac{1}{3}\%$

2. Change the percent to a fraction.

   $33\frac{1}{3}\% = \frac{1}{3}$

3. Multiply 24 by $\frac{1}{3}$ to find the part.

   $\dfrac{\overset{8}{\cancel{24}}}{1} \times \dfrac{1}{\underset{1}{\cancel{3}}} = \dfrac{8}{1} = 8$

Find the part.

1. What is $66\frac{2}{3}\%$ of 90?

   whole = 90
   percent = $66\frac{2}{3}\% = \frac{2}{3}$

   part = $\dfrac{\overset{30}{\cancel{90}}}{1} \times \dfrac{2}{\underset{1}{\cancel{3}}} = \dfrac{60}{1} = 60$

2. What is $83\frac{1}{3}\%$ of 66?

3. What is $16\frac{2}{3}\%$ of 180?

4. What is $83\frac{1}{3}\%$ of 216?

5. What is $16\frac{2}{3}\%$ of 42?

6. What is $33\frac{1}{3}\%$ of 720?

7. A veterinarian gave 99 shots. She gave $66\frac{2}{3}\%$ of the shots to dogs. How many shots did she give to dogs?

8. She gave $33\frac{1}{3}\%$ of the shots to cats. How many shots did she give to cats?

Answer_____

Answer_____

Finding the Part: Percents Greater than 100%

Sometimes the percent you're looking for will be greater than 100%. To solve the problem, you can change the percent to a decimal or to a fraction, whichever is easier. Remember, if you have 100% of something, you have the whole thing.

Use These Steps

What is 110% of 50?

1. Write the pieces.

   whole = 50
   percent = 110%
   part = 50 × 110%

2. Change the percent to a decimal or to an improper fraction. Reduce.

   $110\% = 110. = 1.10 = 1.1$

   or

   $110\% = \dfrac{11\cancel{0}}{10\cancel{0}} = \dfrac{11}{10}$

3. Multiply 50 by 1.1 or by $\dfrac{11}{10}$ to find the part.

   $50 \times 1.1 = 55.\cancel{0} = 55$

   or

   $50 \times \dfrac{11}{10} = \dfrac{\overset{5}{\cancel{50}}}{1} \times \dfrac{11}{\underset{1}{\cancel{10}}} = \dfrac{55}{1} = 55$

Find the part.

1.
What is 250% of 90?
whole = 90
percent = 250% = 2.50 or $\dfrac{5}{2}$
part = 90 × 2.5 = 225

or

$90 \times \dfrac{5}{2} = \dfrac{\overset{45}{\cancel{90}}}{1} \times \dfrac{5}{\underset{1}{\cancel{2}}} = \dfrac{225}{1} = 225$

2.
What is 130% of 10?

3.
What is 325% of 100?

4.
What is 500% of 75?

5.
What is 110% of 500?

6.
What is 450% of 780?

7.
What is 725% of 60?

8.
What is 900% of 15?

Mixed Review

Find the part.

1.

What is 32% of 75?

2.

What is 25% of 20?

3.

What is 10% of 120?

4.

What is 50% of 48?

5.

What is $12\frac{1}{2}$% of 80?

6.

What is $33\frac{1}{3}$% of 6?

7.

What is $6\frac{1}{4}$% of 400?

8.

What is $14\frac{1}{2}$% of 200?

9.

What is $5\frac{3}{4}$% of 400?

10.

What is 10% of 900?

11.

What is 400% of 29?

12.

What is 105% of 700?

13. Brent bought car insurance from Everystate Insurance Company. The total bill was $400. He made a down payment of 20% of the total bill. How much was Brent's down payment?

14. 95% of the students who enter Success Business School will graduate. If 320 students enter the school next year, how many will graduate?

Answer_____

Answer_____

Problem Solving: Using an Interest Formula

You earn interest on money you keep in a savings account. You pay interest on a loan. You can compute simple interest by using the following formula. A formula uses letters which can be replaced with numbers.

The simple interest formula is   I = prt

I = interest
p = principal—the amount borrowed or saved
r = rate—the percent charged each year
   (annually) for loans or earned on savings
t = time in years

I = prt tells you that you find the amount of
   interest by multiplying the principal by the
   rate by the time.

Example   Gary borrowed $600 from the bank. The bank charged $9\frac{1}{2}\%$ annual interest. Gary paid back the money in 1 year. How much interest did he pay?

▶ **Step 1.**   Write the formula. Substitute the given information for each letter.

$$I = prt$$
$$I - \$600 \times 9\frac{1}{2}\% \times 1$$

▶ **Step 2.**   Change the percent to a decimal. Multiply to find the answer.

$$I = \$600 \times .095 \times 1 = \$57.000$$
$$\$57.000 \text{ rounds to } \$57.00$$

Gary paid $57.00 in interest.

Solve. Round to the nearest hundredth, or cent.

1. Sandra borrowed $600 for 1 year. The bank charged $8\frac{1}{2}\%$ annual interest. How much did Sandra pay in interest?

2. Eric borrowed $2,000 for 1 year. The bank charged $10\frac{3}{4}\%$ annual interest. How much did Eric pay in interest?

Answer_____

Answer_____

Solve. Round to the nearest hundredth, or cent.

3. Nancy borrowed $900 from her credit union for 2 years. The annual interest rate for the loan was 7%. How much did she pay in interest charges?

Answer_____

4. Brian is planning to borrow $2,500 from a friend to buy a car. He will pay 6% annual interest and pay back the loan in $2\frac{1}{2}$ years. How much interest will he pay?

Answer_____

5. Lolita kept $200 in her savings account for a year. The bank pays $6\frac{1}{2}\%$ interest per year. How much interest did she receive at the end of 1 year?

Answer_____

6. Lee deposited $275 in his savings account 2 years ago. He earns 4% interest annually. How much interest has he earned?

Answer_____

7. Francis inherited $2,000. She put the money into an account paying $7\frac{1}{4}\%$ annual interest. How much interest did she receive in 1 year?

Answer_____

8. At the beginning of the year, Fern deposited $800 in a savings account paying $6\frac{1}{4}\%$ annual interest. How much interest did she earn in 1 year?

Answer_____

9. Jessie's savings account earns $6\frac{3}{4}\%$ annual interest. If he deposits $400 in his account, how much interest will he receive at the end of 1 year?

Answer_____

10. Bonita had $100 in a savings account paying 5% annual interest. How much interest will she receive at the end of the first year?

Answer_____

Finding the Part: Using a Proportion

You can also solve percent problems by using a proportion. Since percent means hundredths, you can use the following proportion to find the missing number.

$$\frac{part}{whole} = \frac{\%}{100}$$

Use *n* to stand for the part. Cross-multiply to find n, the part.

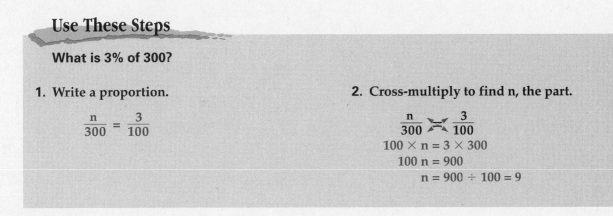

Use These Steps

What is 3% of 300?

1. Write a proportion.

$$\frac{n}{300} = \frac{3}{100}$$

2. Cross-multiply to find n, the part.

$$\frac{n}{300} \diagup\!\!\!\!\diagdown \frac{3}{100}$$

$$100 \times n = 3 \times 300$$
$$100\, n = 900$$
$$n = 900 \div 100 = 9$$

Write a proportion for each problem. Cross-multiply to find the part.

1. What is 30% of 40?

$$\frac{n}{40} \diagup\!\!\!\!\diagdown \frac{30}{100}$$
$$100 \times n = 30 \times 40$$
$$100\, n = 1,200$$
$$n = 1,200 \div 100 = 12$$

2. What is 10% of 90?

3. What is 25% of 32?

4. What is 75% of 80?

5. What is 85% of 100?

6. What is 50% of 10?

7. 52% of the employees at Miller Tire Factory belong to the union. How many of the 125 employees belong to the union?

8. Maxine spends 15% of her grocery budget on paper products. If she spends $120 a week for groceries, how much does she spend for paper products?

Answer_____

Answer_____

Finding the Part: Using a Proportion

When cross-multiplying, remember to multiply n by the number in the opposite corner first.

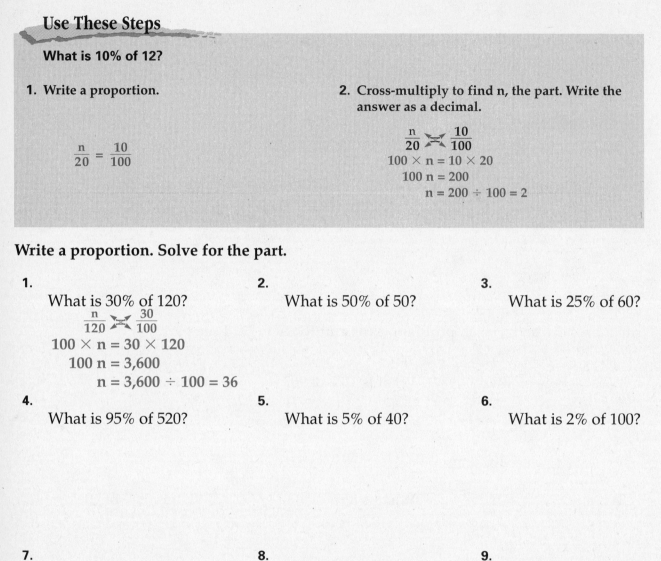

Use These Steps

What is 10% of 12?

1. Write a proportion.

$$\frac{n}{20} = \frac{10}{100}$$

2. Cross-multiply to find n, the part. Write the answer as a decimal.

$$\frac{n}{20} \diagdown\diagup \frac{10}{100}$$
$$100 \times n = 10 \times 20$$
$$100\,n = 200$$
$$n = 200 \div 100 = 2$$

Write a proportion. Solve for the part.

1.

What is 30% of 120?
$$\frac{n}{120} \diagdown\diagup \frac{30}{100}$$
$$100 \times n = 30 \times 120$$
$$100\,n = 3{,}600$$
$$n = 3{,}600 \div 100 = 36$$

2.

What is 50% of 50?

3.

What is 25% of 60?

4.

What is 95% of 520?

5.

What is 5% of 40?

6.

What is 2% of 100?

7.

What is 10% of 750?

8.

What is 20% of 300?

9.

What is 15% of 240?

Problem Solving: Using Percents

When prices go up by a certain percent, you can figure out what the new price will be. The new price will be 100% of the old price plus a percent of the old price.

Example The price of a truck has increased by 10%. If the old price was $6,590, what is the new price?

▶ **Step 1.** The new price is 100% of the old price plus 10% of the old price.

$$100\% + 10\% = 110\%$$

▶ **Step 2.** Write a proportion. Cross-multiply.

$$\frac{n}{\$6,590} \diagdown\!\!\!\diagup \frac{110}{100}$$

$$n \times 100 = \$6,590 \times 110$$
$$100\,n = \$724,900$$
$$n = \$724,900 \div 100 = \$7,249$$

The new price is $7,249.

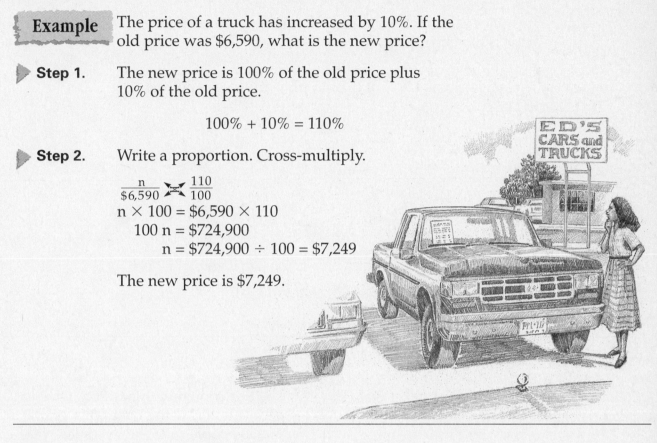

Solve.

1. Last year Peter made $15,000. This year he received a 5% raise. How much will he make this year?

   Answer_____

2. Last year the Lasher Corporation employed 40 people. This year they have 20% more people working for them. How many people now work at Lasher's?

   Answer_____

3. Ten years ago, Sklar Systems produced 60,000 stereo systems. This year the company increased production by 15%. How many stereo systems did the company produce this year?

   Answer_____

4. Before its last tune-up, Hal's truck went 15 miles on a gallon of gas. After the tune-up it went 20% farther. How many miles per gallon did the truck get after the tune-up?

   Answer_____

When prices go down by a certain percent, you can figure out the new price. The new price will be 100% of the old price minus a percent of the old price.

Example | The local used car dealership is having a 25% off clearance sale on certain cars. If a car used to cost $5,000, what is the new price?

Step 1. The new price is 100% of the old price minus 25% of the old price.

$$100\% - 25\% = 75\%$$

Step 2. Write a proportion. Cross-multiply.

$$\frac{n}{\$5,000} \quad \frac{75}{100}$$

$$n \times 100 = \$5,000 \times 75$$
$$100\,n = \$375,000$$
$$n = \$375,000 \div 100 = \$3,750$$

The new price is $3,750.

Solve.

1. Last year the price of a bus ticket was $4. This year the price decreased by 25%. What is the new price?

    Answer_____

2. The bus company thinks 20% fewer people will ride the buses this year. If 24,250 people rode the buses last year, how many people will ride the buses this year?

    Answer_____

3. Last year Athena worked 40 hours per week. This year she will work 30% fewer hours. How many hours per week will she work?

    Answer_____

4. Last year Athena's take-home pay was $200 per week. This year she will be taking home 25% less money. What will her weekly pay be?

    Answer_____

Finding the Part: Using a Proportion With Complex Percents

To use a proportion to solve problems with complex percents, use the fraction equivalent for the percent. Then cross-multiply.

You can refer to the table on page 94 for some common percents and their fraction equivalents.

Use These Steps

What is $33\frac{1}{3}$% of 150?

1. Write a proportion. Write $33\frac{1}{3}$% as $\frac{1}{3}$.

$$\frac{n}{150} = \frac{1}{3}$$

2. Cross-multiply to find n, the part.

$$\frac{n}{150} \diagup\!\!\!\!\diagdown \frac{1}{3}$$
$$n \times 3 = 150 \times 1$$
$$3\,n = 150$$
$$n = 150 \div 3 = 50$$

Use the table on page 94 to find the equivalent fractions. Write a proportion. Find the part.

1. What is $66\frac{2}{3}$% of 90?
$$\frac{n}{90} \diagup\!\!\!\!\diagdown \frac{2}{3}$$
$$n \times 3 = 90 \times 2$$
$$3\,n = 180$$
$$n = 180 \div 3 = 60$$

2. What is $12\frac{1}{2}$% of 16?

3. What is $62\frac{1}{2}$% of 32?

4. What is $16\frac{2}{3}$% of 6?

5. What is $37\frac{1}{2}$% of 64?

6. What is $83\frac{1}{3}$% of 240?

7. Associated Airlines offers senior citizens a $33\frac{1}{3}$% discount on airfare. How much would a senior citizen save on a $450 ticket to Florida?

8. $37\frac{1}{2}$% of the 88 movies on television last week were comedies. How many movies were comedies?

Answer_____

Answer_____

Real-Life Application

At the Store

When you figure the amount of sales tax you pay or the amount you save at a sale, you are finding the part.

Example The sales tax in Johnson City is 7%. If Sue buys a $200 couch in Johnson City, how much sales tax will she pay?

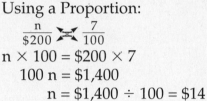

Using a Percent Sentence:

part (tax) = whole (cost) × percent (tax rate)

part = $200 × 7%
part = $200 × .07 = $14

Sue will pay $14 in tax.

Using a Proportion:

$$\frac{n}{\$200} \times \frac{7}{100}$$

n × 100 = $200 × 7
100 n = $1,400
n = $1,400 ÷ 100 = $14

Solve. Use either a percent sentence or a proportion.

1. Mary's Clothing Mart is having a sale. Everything is 25% off. How much will you save on a $40 dress?

   Answer_____

2. Book Barn sells books at $33\frac{1}{3}$% off the regular price. Vince bought a book that usually sells for $30. How much did he save at Book Barn?

   Answer_____

3. Place Appliance is offering a discount of 40% on a discontinued washing machine. If the washer normally costs $350, how much is the discount?

   Answer_____

4. Allen spent 10% of his monthly pay on a new television. He earned $1,520 in April. How much did he spend on the television?

   Answer_____

5. Last year the price of a Handy home computer was $1,200. This year the price decreased 32%. How much less does the computer cost this year?

   Answer_____

6. The sales tax in Landsburg is 8%. How much sales tax did John pay on a $450 stereo he bought there?

   Answer_____

Mixed Review

Find the part using a proportion.

1. What is 20% of 40?

2. What is 50% of 90?

3. What is 37% of 100?

4. What is 60% of 340?

5. What is 4% of 50?

6. What is 1% of 500?

7. What is $33\frac{1}{3}$% of 126?

8. What is $12\frac{1}{2}$% of 256?

9. What is $16\frac{2}{3}$% of 90?

10. What is $66\frac{2}{3}$% of 72?

11. What is 300% of 10?

12. What is 130% of 280?

13. Cora put $85 worth of toys on lay away. She had to make a 20% down payment. How much was Cora's down payment?

14. Thriftee Cleaners gives a 15% discount on cleaning orders over twenty dollars. Tina's total order was $40. How much money did she save?

Answer _____

Answer _____

Finding the Whole: Changing Percents to Decimals

You can find the whole in a percent problem by using this sentence:

whole = part ÷ percent

Change the percent to a decimal. Then divide.

Use These Steps

22 is 10% of what number?

1. Write the pieces.

part = 22
percent = 10%
whole = 22 ÷ 10%

2. Change the percent to a decimal.

10% = .1

3. Divide 22 by .1 to find the whole.

22 ÷ .1 = 220

Find the whole.

1.
38 is 20% of what number?
part = 38
percent = 20% = .2
whole = 38 ÷ .2 = 190

2.
50 is 200% of what number?

3.
7 is 5% of what number?

4.
15 is 3% of what number?

5.
12 is 50% of what number?

6.
90 is 45% of what number?

7.
120 is 60% of what number?

8.
420 is 70% of what number?

Finding the Whole: Changing Complex Percents

In some problems, the percent may be a complex percent. To solve the problem, change the percent to a decimal.

Use These Steps

11 is $5\frac{1}{2}$% of what number?

1. Write the pieces.

    part = 20
    percent = $5\frac{1}{2}$%
    whole = 11 ÷ $5\frac{1}{2}$%

2. Change the percent to a decimal.

    $5\frac{1}{2}$% = 5.5% = .055

3. Divide 11 by .055 to find the whole.

    11 ÷ .055 = 200

Solve for the part.

1. 36 is $37\frac{1}{2}$% of what number?
   part = 36
   percent = $37\frac{1}{2}$% = .375
   whole = 36 ÷ .375 = 96

2. 45 is $62\frac{1}{2}$% of what number?

3. 97 is $24\frac{1}{4}$% of what number?

4. 594 is $6\frac{3}{4}$% of what number?

5. 22 is $4\frac{2}{5}$% of what number?

6. 3 is $1\frac{1}{2}$% of what number?

7. 4 is $12\frac{1}{2}$% of what number?

8. 21 is $5\frac{1}{4}$% of what number?

9. 175 is $87\frac{1}{2}$% of what number?

10. 17 is $8\frac{1}{2}$% of what number?

Finding the Whole: Changing Percents to Fractions

Sometimes it is easier to change the percent to a fraction instead of to a decimal. Remember, when you divide by a fraction, invert the fraction and then multiply. Use the table on page 94 to help you.

Use These Steps

50 is 25% of what number?

1. Write the pieces.

    part = 50
    percent = 25%
    whole = 50 ÷ 25%

2. Change the percent to a fraction. Reduce if possible.

    $$25\% = \frac{25}{100} = \frac{1}{4}$$

3. Divide 50 by $\frac{1}{4}$ to find the whole.

    $$\frac{50}{1} \div \frac{1}{4} = \frac{50}{1} \times \frac{4}{1} = \frac{200}{1} = 200$$

Find the whole.

1.

70 is 50% of what number?

part = 70

$$\text{percent} = 50\% = \frac{50}{100} = \frac{1}{2}$$

$$\text{whole} = \frac{70}{1} \div \frac{1}{2} = \frac{70}{1} \times \frac{2}{1} = \frac{140}{1} = 140$$

2.

32 is 20% of what number?

3.

20 is 250% of what number?

4.

10 is 40% of what number?

5.

4 is 10% of what number?

6.

88 is 22% of what number?

7. 75% of the registered voters in Met County voted in May. If 1,500 people voted, how many registered voters are there in Met County?

8. 20% of the units in Hill Apartments are two-bedroom units. If 12 units have two bedrooms, how many units are there in the whole apartment complex?

Answer_____

Answer_____

Finding the Whole: Complex Percents

Some problems use complex percents such as $33\frac{1}{3}\%$ and $66\frac{2}{3}\%$. When this happens, change the percent to a fraction instead of to a decimal. Then invert and multiply. You may want to use the table on page 94 to help you.

Use These Steps

50 is $66\frac{2}{3}\%$ of what number?

1. Write the pieces.

part = 50
percent = $66\frac{2}{3}\%$
whole = $50 \div 66\frac{2}{3}\%$

2. Change the percent to a fraction.

$$66\frac{2}{3}\% = \frac{2}{3}$$

3. Divide 50 by $\frac{2}{3}$ to find the whole.

$$\frac{50}{1} \div \frac{2}{3} = \frac{\overset{25}{\cancel{50}}}{1} \times \frac{3}{\underset{1}{\cancel{2}}} = \frac{75}{1} = 75$$

Find the whole.

1.

18 is $12\frac{1}{2}\%$ of what number?

part = 18

percent = $12\frac{1}{2}\% = \frac{1}{8}$

whole = $\frac{18}{1} \div \frac{1}{8} = \frac{18}{1} \times \frac{8}{1} = \frac{144}{1} = 144$

2.

25 is $62\frac{1}{2}\%$ of what number?

3.

95 is $33\frac{1}{3}\%$ of what number?

4.

102 is $16\frac{2}{3}\%$ of what number?

5.

8 is $66\frac{2}{3}\%$ of what number?

6.

12 is $37\frac{1}{2}\%$ of what number?

7.

8 is $16\frac{2}{3}\%$ of what number?

8.

15 is $83\frac{1}{3}\%$ of what number?

9.

320 is $66\frac{2}{3}\%$ of what number?

10.

25 is $33\frac{1}{3}\%$ of what number?

Mixed Review

Find the missing piece.

1.

What is 50% of 10?

2.

16 is 20% of what number?

3.

What is 1% of 250?

4.

What is 30% of 90?

5.

25 is $33\frac{1}{3}$% of what number?

6.

90 is 300% of what number?

7.

What is 150% of 40?

8.

2 is 10% of what number?

9.

16 is 8% of what number?

10.

10 is 5% of what number?

11.

What is $12\frac{1}{2}$% of 160?

12.

What is 17% of 600?

Finding the Whole: Using a Proportion

When the missing number is the whole, you can find it by using a proportion. Use the proportion:

$$\frac{\text{part}}{\text{whole}} = \frac{\%}{100}$$

Use These Steps

6 is 5% of what number?

1. Write a proportion.

$$\frac{6}{n} = \frac{5}{100}$$

2. Cross-multiply to find n, the whole.

$$\frac{6}{n} \diagup\kern-1em\diagdown \frac{5}{100}$$

$n \times 5 = 6 \times 100$

$5\,n = 600$

$n = 600 \div 5 = 120$

Write a proportion for each problem. Cross-multiply to find the whole.

1.

10 is 40% of what number?

$$\frac{10}{n} \diagup\kern-1em\diagdown \frac{40}{100}$$

$n \times 40 = 10 \times 100$

$40\,n = 1,000$

$n = 1,000 \div 40 = 25$

2.

39 is 1% of what number?

3.

80 is 10% of what number?

4.

66 is 120% of what number?

5.

30 is 75% of what number?

6.

47 is 5% of what number?

Finding the Whole: Using a Proportion

In problems with complex percents, change the percent to an equivalent fraction. Then cross-multiply to find the missing number.

Use These Steps

18 is $33\frac{1}{3}$% of what number?

1. Write a proportion. Write $33\frac{1}{3}$% as $\frac{1}{3}$.

$$\frac{18}{n} = \frac{1}{3}$$

2. Cross-multiply to find n, the whole.

$$\frac{18}{n} \diagup\!\!\!\!\diagdown \frac{1}{3}$$

$$1 \times n = 18 \times 3$$

$$n = 54$$

Write a proportion for each problem. Cross-multiply to find the whole.

1.

20 is $66\frac{2}{3}$% of what number?

$$\frac{20}{n} \diagup\!\!\!\!\diagdown \frac{2}{3}$$

$$n \times 2 = 20 \times 3$$

$$2n = 60$$

$$n = 60 \div 2 = 30$$

2.

6 is $12\frac{1}{2}$% of what number?

3.

45 is $83\frac{1}{3}$% of what number?

4.

2 is $16\frac{2}{3}$% of what number?

5.

87 is $37\frac{1}{2}$% of what number?

6.

110 is $33\frac{1}{3}$% of what number?

7.

48 is $66\frac{2}{3}$% of what number?

8.

30 is $62\frac{1}{2}$% of what number?

Real-Life Application   At the Store

In many situations, you know the percent and the part, and you need to find the original amount.

Example Mark's Resale Shop is having a 25%-off sale. Joe saved $10 on a winter coat. What was the original cost of the coat?

$$\frac{\text{part (savings)}}{\text{whole (original cost)}} = \frac{\text{percent (sale \%)}}{100}$$

$$\frac{\$10}{n} \bowtie \frac{25}{100}$$

$$n \times 25 = \$10 \times 100$$
$$25\,n = \$1{,}000$$
$$n = \$1{,}000 \div 25 = \$40$$

The coat originally cost $40.

Solve.

1. Value City sells radios at a 20% discount. Frank saved $4 on a radio. What was the original cost?

   Answer_____

2. Collier's Appliance is offering a 10% discount on the cost of a certain refrigerator. If the Petersons saved $35, what was the original cost?

   Answer_____

3. Lisa saves 15% of her monthly pay. She saved $150 last month. How much does Lisa make a month?

   Answer_____

4. Thomas saved 25% because he brought his car in for an oil change before 10 a.m. He saved $7. How much would he have paid after 10 a.m.?

   Answer_____

5. A car dealer is offering a rebate of $600. The dealer says that this amount is 10% of the price of a certain car. How much is the car?

   Answer_____

6. In the Southwest Mall, 75% of the stores are open for business. Cathy counted 30 open stores. How many stores are there in the mall?

   Answer_____

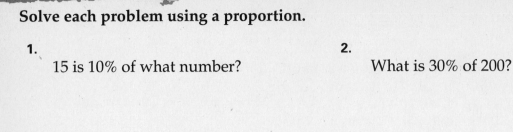

Mixed Review

Solve each problem using a proportion.

1. 15 is 10% of what number?

2. What is 30% of 200?

3. What is 20% of 45?

4. 87 is 25% of what number?

5. 44 is 110% of what number?

6. What is 95% of 80?

7. What is 50% of 10?

8. 6 is 75% of what number?

9. 10 is $33\frac{1}{3}$% of what number?

10. What is $87\frac{1}{2}$% of 16?

11. What is $66\frac{2}{3}$% of 12?

12. 3 is 15% of what number?

Finding the Percent

You can find the percent in a percent problem by using this percent sentence:

$$percent = part \div whole$$

Use These Steps

30 is what percent of 120?

1. Write the pieces.

part = 30
whole = 120
percent = 30 ÷ 120

2. Divide 30 by 120.

30 ÷ 120 = .25

3. Change .25 to a percent.

.25 = 25%

Find the percent.

1.
10 is what percent of 20?
part = 10
whole = 20
percent = 10 ÷ 20 = .5 = 50%

2.
27 is what percent of 108?

3.
8 is what percent of 40?

4.
60 is what percent of 600?

5.
93 is what percent of 100?

6.
146 is what percent of 2,920?

7.
340 is what percent of 400?

8.
120 is what percent of 300?

9.
430 is what percent of 500?

10.
110 is what percent of 250?

Finding the Percent

When finding the percent, you can write the percent as a decimal or as its fraction equivalent.

Use These Steps

15 is what percent of 1,000?

1. Write the pieces.

part = 15
whole = 1,000
percent = 15 ÷ 1,000

2. Divide 15 by 1,000.

15 ÷ 1,000 = .015

3. Change .015 to a percent.

$.015 = 1.5\%$ or $1\frac{1}{2}\%$

Find the percent.

1.
8 is what percent of 320?
part = 8
whole = 320
percent = $8 \div 320 = .025 = 2.5\%$ or $2\frac{1}{2}\%$

2.
10 is what percent of 400?

3.
19 is what percent of 500?

4.
30 is what percent of 800?

5.
75 is what percent of 6,000?

6.
84 is what percent of 960?

7.
6 is what percent of 250?

8.
50 is what percent of 2,000?

9.
24 is what percent of 64?

10.
400 is what percent of 640?

11.
210 is what percent of 240?

12.
189 is what percent of 1,500?

Finding the Percent

In some problems, the answer will be a complex percent.

Use These Steps

25 is what percent of 75?

1. Write the pieces.

   part = 25
   whole = 75
   percent = 25 ÷ 75

2. Divide 25 by 75.

   $25 \div 75 = .33\frac{1}{3}$

3. Change $.33\frac{1}{3}$ to a percent.

   $.33\frac{1}{3} = 33\frac{1}{3}\%$

Find the percent.

1.
   20 is what percent of 60?
   part = 20
   whole = 60
   percent = $20 \div 60 = 33\frac{1}{3} = 33\frac{1}{3}\%$

2. 14 is what percent of 21?

3. 5 is what percent of 60?

4. 16 is what percent of 24?

5. 11 is what percent of 66?

6. 85 is what percent of 102?

7. 22 is what percent of 66?

8. 96 is what percent of 144?

9. 75 is what percent of 90?

10. 12 is what percent of 72?

Mixed Review

Find the percent.

1. 4 is what percent of 8?

2. 22 is what percent of 88?

3. 9 is what percent of 120?

4. 15 is what percent of 200?

5. 38 is what percent of 228?

6. 40 is what percent of 60?

7. 93 is what percent of 100?

8. 2 is what percent of 50?

9. 125 is what percent of 625?

10. 50 is what percent of 5,000?

11. 70 is what percent of 280?

12. 85 is what percent of 136?

13. 31 is what percent of 93?

14. 65 is what percent of 78?

Finding the Percent: Using a Proportion

You can use a proportion to find the percent. Use the proportion:

$$\frac{\text{part}}{\text{whole}} = \frac{\%}{100}$$

Substitute the correct numbers in the proportion and cross-multiply to find the percent.

Notice when you use a proportion, you write a percent sign after the answer.

Use These Steps

5 is what percent of 25?

1. Write the proportion.

$$\frac{5}{25} = \frac{n}{100}$$

2. Cross-multiply to find n. Add a percent sign.

$$\frac{5}{25} \diagup\!\!\!\!\diagdown \frac{n}{100}$$
$$25 \times n = 5 \times 100$$
$$25\,n = 500$$
$$n = 500 \div 25 = 20 = 20\%$$

Write a proportion for each problem. Cross-multiply to find the percent.

1.

10 is what percent of 200?

$$\frac{10}{200} \diagup\!\!\!\!\diagdown \frac{n}{100}$$
$$200 \times n = 10 \times 100$$
$$200\,n = 1,000$$
$$n = 1,000 \div 200 = 5 = 5\%$$

2.

8 is what percent of 64?

3.

15 is what percent of 45?

4.

33 is what percent of 55?

5.

70 is what percent of 100?

6.

63 is what percent of 420?

Percent of Increase

When prices increase, people often want to know what percent this increase represents. To find the percent, subtract the old price from the new price. Divide the answer by the old price. Change the answer to a percent.

Use These Steps

What is the percent of increase from $3 to $4?

1. Subtract the old price from the new price to find the change in price.

    $$\$4 - \$3 = \$1$$

2. Divide the change in price by the old price.

    $$\$1 \div \$3 = .33\frac{1}{3}$$

3. Change to a percent.

    $$.33\frac{1}{3} = 33\frac{1}{3}\%$$

Find the percent of increase.

	old price	new price	change in price	% of increase
1.	$ 2	$ 3	$3 − $2 = $1	$1 ÷ $2 = .5 = 50%
2.	$ 8	$ 10		
3.	$ 10	$ 13		
4.	$ 20	$ 22		
5.	$ 32	$ 40		
6.	$ 36	$ 72		
7.	$ 80	$100		
8.	$150	$175		

Percent of Decrease

Sometimes the prices on some items go down. To find the percent of decrease, subtract the new price from the old price. Then divide by the old price. Change the answer to a percent.

Use These Steps

What is the percent of decrease from $20 to $15?

1. Subtract the new price from the old price to find the change in price.

   $20 − $15 = $5

2. Divide the change in price by the old price.

   $5 ÷ $20 = .25

3. Change to a percent.

   .25 = 25%

Find the percent of decrease.

	old price	new price	change in price	% of decrease
1.	$ 50	$ 45	$50 − $45 = $5	$5 ÷ $50 = .1 = 10%
2.	$ 20	$ 15		
3.	$120	$110		
4.	$ 36	$ 30		
5.	$ 75	$ 60		
6.	$ 10	$ 8		
7.	$100	$ 50		
8.	$ 6	$ 4		

Unit 5 *Review*

Find the part using a percent sentence.

1.

   What is 27% of 200?

2.

   What is $12\frac{1}{2}$% of 80?

3.

   What is $2\frac{1}{2}$% of 400?

4.

   What is 110% of 90?

Find the part using a proportion.

5.

   What is 30% of 50?

6.

   What is 25% of 80?

7.

   What is 5% of 60?

8.

   What is $33\frac{1}{3}$% of 60?

Find the whole using a percent sentence.

9.

   15 is 3% of what number?

10.

   21 is 14% of what number?

11.

   8 is $12\frac{1}{2}$% of what number?

12.

   44 is $4\frac{2}{5}$% of what number?

Find the whole using a proportion.

13.

45 is 50% of what number?

14.

9 is 300% of what number?

15.

90 is $33\frac{1}{3}$% of what number?

16.

40 is $66\frac{2}{3}$% of what number?

Find the percent using a percent sentence.

17.

19 is what percent of 500?

18.

27 is what percent of 108?

19.

25 is what percent of 150?

20.

21 is what percent of 240?

Find the percent using a proportion.

21.

13 is what percent of 26?

22.

120 is what percent of 1,200?

23.

80 is what percent of 320?

24.

6 is what percent of 30?

Find the percent of increase or decrease.

	old	new	change in price	% of increase or decrease
25.	$ 50	$ 55		
26.	$ 30	$ 25		
27.	$ 16	$ 18		
28.	$ 5	$ 4		
29.	$ 25	$ 15		
30.	$ 60	$ 70		
31.	$150	$120		
32.	$ 40	$ 30		
33.	$180	$240		
34.	$ 75	$150		

Below is a list of the problems in this review and the pages on which the skills are taught. If you missed any problems, turn to the pages listed and practice the skills. Then correct the problems you missed in the Unit Review.

Problems	Pages	Problems	Pages
1-4	107-111	17-20	131-133
5-8	115-116, 119	21-24	135
9-12	122-125	25-34	136-137
13-16	127-128		

You have studied decimals, ratios, proportions, and percents. You have applied these skills to real-life problems, and you have learned techniques for solving word problems.

In this unit, you will learn how to choose the correct operation needed to solve problems. You will also learn how to solve problems that require you to use more than one operation to find the answers.

Getting Ready

You should be familiar with the skills on this page and the next before you begin this unit. To check your answers, turn to page 195.

▶ When adding and subtracting decimals, line up the decimal points. Add zeros and decimal points if necessary.

Add or subtract.

1.
$1.2 + 4.72 =$
$$\begin{array}{r} 1.20 \\ + 4.72 \\ \hline 5.92 \end{array}$$

2.
$12.6 + .02 =$

3.
$213.7 + 6.98 =$

4.
$85.06 + .04 =$

5.
$947.6 - 82.9 =$

6.
$14 - 1.03 =$

7.
$72.3 - 41.91 =$

8.
$493.1 - .036$

When multiplying and dividing decimals, be sure to include a decimal point in the answer.

Multiply or divide.

9.

$1.3 \times 3 =$

$$\begin{array}{r} 1.3 \\ \times\ \ 3 \\ \hline 3.9 \end{array}$$

10.

$.03 \times .07 =$

11.

$12.72 \div 8 =$

12.

$71.04 \div 32 =$

For review, see Unit 3.

Sometimes it is necessary to change a percent to a decimal or a fraction before solving a problem.

Change each percent to a decimal and a fraction. Reduce if possible.

13.

$50\% = .50 = \frac{1}{2}$

14.

$2\% =$

15.

$25\% =$

16.

$33\frac{1}{3}\% =$

17.

$670\% =$

18.

$16\frac{2}{3}\% =$

For review, see Unit 4.

You can find the missing piece in a percent problem by using a percent sentence: part = whole × percent.

Find the part, the whole, and the percent.

19.

What is 50% of 40?
whole = 40
percent = 50% = .5
part = 40 × .5 = 20

20.

What is $33\frac{1}{3}\%$ of 90?

21.

16 is 25% of what?

22.

100 is what percent of 500?

For review, see Unit 5.

Choose an Operation: Being a Consumer

If you have a savings account, you work with decimals all the time. Each time you make a deposit, you add the amount of the deposit to your balance. Each time you withdraw money from your account, you subtract the amount of the withdrawal from your balance. And when the bank pays you interest, you add that amount to your balance.

Example On January 1, the balance in Vanessa's savings account was $356.89. On payday, January 15, she deposited $37. What was her new balance after she made the deposit?

▶ **Step 1.** To find her new balance, add her deposit to her old balance from January 1.

$356.89
+   37.00
$393.89

Date	Withdrawal		Deposit		Interest Credited		Balance	
Jan. 1							356	89
Jan. 15			37	00			393	89
Jan. 31								
Feb. 10	188	23					207	30
Feb. 15								
Feb. 16			450	00			581	80
Feb. 27	525	77					56	03
Feb. 28								
March 15			442	01			500	00
March 21								

▶ **Step 2.** Write her new balance in the last column on the same line in her bankbook as her deposit.

Vanessa's balance on January 15 was $393.89.

Find the balance in Vanessa's savings account.

1. On January 31, the bank paid Vanessa $1.64 in interest. What was her new balance on January 31?

Answer_____

2. On February 15 Vanessa had a balance of $207.30. She took out $75.50. What was her new balance?

Answer_____

3. On February 28, Vanessa had a balance of $56.03. The bank paid Vanessa $1.96 in interest. Circle the expression you would use to find Vanessa's new balance.
   a. $56.03 − $1.96   Solve for the
   b. $56.03 + $1.96   answer.
   c. $56.03 × $1.96
   d. $56.03 ÷ $1.96

Answer_____

4. On March 21, Vanessa had a balance of $500.00. She took out $234.75. Circle the expression you would use to find Vanessa's new balance.
   a. $500.00 − $234.75   Solve for the
   b. $500.00 + $234.75   answer.
   c. $500.00 × $234.75
   d. $500.00 ÷ $234.75

Answer_____

Choose an Operation: Using Data Analysis

After a basketball game, newspapers list the top players, the number of games they played, and the number of points they scored. To find a player's average score, sports writers divide the number of points the player scored by the number of games the player played. They round the answer to the nearest tenth.

Example Gordon has played in 44 games this season. He has scored 1,305 points. How many points did Gordon score per game?

▶ **Step 1.** $1,305 \div 44 = 29.65$

▶ **Step 2.** 29.65 rounded to the nearest tenth is 29.7.

Gordon averaged 29.7 points per game.

Solve.

1. Williams played in 42 games. He scored 1,192 points. What is his average score in a game? Round the answer to the nearest tenth.

2. Maulins played in 39 games. He averaged 26.9 points per game. How many total points did he score? Round the answer to the nearest whole number.

Answer_____

Answer_____

3. Wing played in 42 games. He scored 1,003 points. Circle the expression you would use to find his average score. Round the answer to the nearest tenth.
   a. 1,003 + 42   Solve for the answer.
   b. 1,003 − 42
   c. 1,003 × 42
   d. 1,003 ÷ 42

4. Hardy played in 39 games. He averaged 23.4 points per game. Circle the expression you would use to find the total points he scored. Round the answer to the nearest whole number.
   a. 23.4 + 39   Solve for the answer.
   b. 23.4 − 39
   c. 23.4 × 39
   d. 23.4 ÷ 39

Answer_____

Answer_____

Choose an Operation: Changing Units of Measurement

Sometimes when we measure, we use the metric system. In the metric system, weight is measured in grams and length is measured in meters. 454 grams is equal to about 1 pound. One meter is equal to about 39 inches. Grams and meters are base units. Large weights are measured in kilograms. Long distances are measured in kilometers.

To change from one unit of measurement to another, remember that when you change from a small unit to a larger unit, you divide. When you change from a large unit to a smaller unit, you multiply.

| 1,000 meters = 1 kilometer |
| 1,000 grams = 1 kilogram |

Example Robert bought 1,200 grams of fertilizer for his garden. How many kilograms are in 1,200 grams?

▶ **Step 1.** You are changing from a small unit (grams) to a larger unit (kilograms). You will need to divide.

▶ **Step 2.** Use the chart to find out how many grams are in one kilogram.

1,000 grams = 1 kilogram

▶ **Step 3.** Divide.

1,200 ÷ 1,000 = 1,200 = 1.2

There are 1.2 kilograms in 1,200 grams.

Solve.

1. A large box of detergent weighs .3 kilograms. How many grams are in .3 kilograms?

   Answer_____

2. Edward lives 2.1 kilometers from the nearest gas station. How many meters are in 2.1 kilometers?

   Answer_____

3. Estella swims 500 meters each morning. Circle the expression you would use to find how many kilometers she swims.
   a. 500 × 1,000   Solve for the answer.
   b. 500 − 1,000
   c. 500 ÷ 1,000
   d. 500 + 1,000

   Answer_____

4. A large box of wood chips weighs 7,352 grams. Circle the expression you would use to find how many kilograms are in 7,352 grams.
   a. 7,352 × 1,000   Solve for the answer.
   b. 7,352 − 1,000
   c. 7,352 ÷ 1,000
   d. 7,352 + 1,000

   Answer_____

Multi-Step Problems: Being a Consumer

When you have a credit card, the amount of any purchases you made is added to your statement at the end of each billing period. If you have an unpaid balance, that amount and a finance charge are also added. At the same time, payments you make are credited to your account.

The interest rate on Sheila's Vista charge card is 1.5% each month.

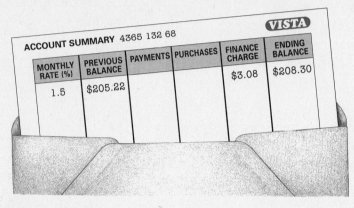

> **Example** Sheila had a previous balance on her Vista charge card of $205.22. She was charged a 1.5% interest rate for the month. What was her ending balance?

▶ **Step 1.** Find the interest by multiplying the previous balance by the monthly interest rate. Change the percent to a decimal before multiplying.

$$1.5\% = .015$$
$$\$205.22 \times .015 = \$3.07830$$
$$\$3.07830 \text{ rounds to } \$3.08$$

▶ **Step 2.** Add the amount of interest to her previous balance.

$$\begin{array}{r} \$205.22 \\ +\quad 3.08 \\ \hline \$208.30 \end{array}$$

Sheila's ending balance was $208.30.

Solve each problem. Round to the nearest hundredth, or cent.

1. Chris had a previous balance of $235.56 on her Stacy's charge card. The interest rate on that card is 1.65% per month. What will her ending balance be?

2. The finance charge on Gil's Maximum Card is 1.5% per month. In August Gil had a previous balance of $155.43. What was his ending balance?

Answer_____

Answer_____

Solve each problem. Round to the nearest hundredth, or cent.

3. James made a payment on his credit card of $100 which included a finance charge of $24.56. What percent of his payment went toward the finance charge?
(Hint: percent = part ÷ whole)

Answer_____

4. The remainder of James' $100 payment went to pay his unpaid balance. What percent of his payment went to pay off his unpaid balance? (Hint: Subtract the finance charge from James' payments.)

Answer_____

5. Edna has a previous balance of $150.56. She makes a payment of $10 and buys a pair of shoes for $29.00. How much will her ending balance be?

Answer_____

6. Later in the year Edna has a previous balance of $143.02. The interest rate is 1.75%. How much will her ending balance be?

Answer_____

7. Edna made a payment of $10 which included a finance charge of $3.89. What percent of her payment went for the finance charge?

Answer_____

8. What percent of Edna's payment went to pay off her balance?

Answer_____

9. Yianna had an ending balance of $120.45. She paid 15% of the ending balance. How much did she pay?
(Hint: part = whole × percent)

Answer_____

10. Troy has an ending balance of $74.00. He must pay either 10% of the ending balance or $20. How much greater is $20.00 than 10% of his balance?

Answer_____

Multi-Step Problems: Using Measurement

You can figure out how far you have gone (distance) if you know how fast you are going (rate) and how long it will take you to get from one place to another (time). The distance formula below shows you how.

distance (D) = rate (R) × time (T)

If you want to find the rate, use this formula:

rate (R) = distance (D) ÷ time (T)

If you want to find the time, use this formula:

time (T) = distance (D) ÷ rate (R)

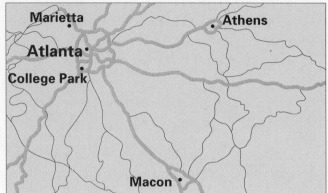

Always show time as a decimal or as a fraction of an hour.

Example How long will it take to drive 275 miles if you travel an average of 50 miles per hour?

Step 1. Since you are looking for time, use the time formula.

$$T = D ÷ R$$

Step 2. Substitute the numbers in the formula. Solve by dividing.

$$T = 275 ÷ 50 = 5.5 \text{ or } 5\frac{1}{2} \text{ hours}$$

It will take 5.5 or $5\frac{1}{2}$ hours.

Solve.

1. Carlos is driving to a new construction site. He needs to travel 30 miles in 45 minutes. How many miles per hour will he travel?

2. The speed limit on one stretch of Sunset Highway is 60 miles per hour. How far can you travel at this rate in 4 hours and 30 minutes?

Answer_____   Answer_____

Solve.

3. April and Martin are traveling 300 miles to Atlanta. How long will it take them to get to Atlanta if they drive 60 miles per hour?

Answer_____

4. April found out that she and Martin can fly to Atlanta in 45 minutes. How many miles per hour will the plane travel?

Answer_____

5. Martin found out that a bus averages 50 miles per hour. If they take the bus, how much longer will it take April and Martin to get to Atlanta than if they take a plane? Write the answer using hours.

Answer_____

6. Normally Yang can make the 75 mile trip to Houston in 1 hour and 15 minutes. What is his usual rate of speed for the trip?

Answer_____

7. On his last trip to Houston, Yang had to take a detour. He averaged 30 miles per hour. How long did it take him to drive the 75 miles to Houston? Write the answer using hours.

Answer_____

8. The speed limit on the detour was 35 miles per hour. Yang drove for 30 minutes at this rate. How long was the detour?

Answer_____

9. The speed limit on Scenic Highway is 55 miles per hour. Next year the speed limit will be raised to 65 miles per hour. How much farther will you be able to travel in 2 hours next year?

Answer_____

10. Due to construction, the speed limit on a stretch of Scenic Highway has been reduced to 30 miles per hour. How far can you travel in 2 hours?

Answer_____

Multi-Step Problems: Using Data Analysis

The federal government spends about 1,252 billion dollars yearly. The circle graph shows what percent of the total amount is spent in each area.

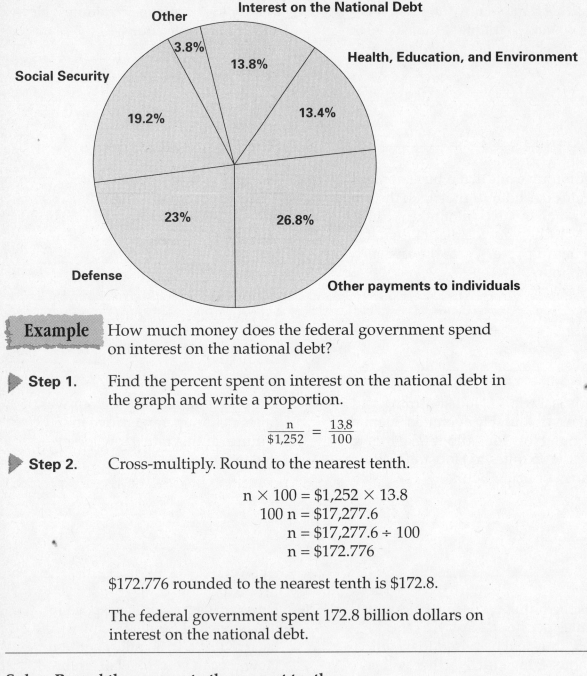

Example   How much money does the federal government spend on interest on the national debt?

▶ **Step 1.**   Find the percent spent on interest on the national debt in the graph and write a proportion.

$$\frac{n}{\$1,252} = \frac{13.8}{100}$$

▶ **Step 2.**   Cross-multiply. Round to the nearest tenth.

$$n \times 100 = \$1,252 \times 13.8$$
$$100\,n = \$17,277.6$$
$$n = \$17,277.6 \div 100$$
$$n = \$172.776$$

$172.776 rounded to the nearest tenth is $172.8.

The federal government spent 172.8 billion dollars on interest on the national debt.

Solve. Round the answer to the nearest tenth.

1. How much money does the federal government spend on Social Security?

2. How much money does the federal government spend on defense?

Answer _____

Answer _____

3. How much money does the federal government spend on health, education, and the environment?

Answer_____

4. How much money does the federal government spend on other payments to individuals?

Answer_____

5. The federal government spends 1% of the budget on science, space, and technology. How much money is 1% of the federal budget?

Answer_____

6. Veterans benefits are 2.3% of the total budget. How much money does the federal government spend on veterans benefits?

Answer_____

7. The federal government spends 150 billion dollars on welfare payments. What percent of the total budget is this?

Answer_____

8. The federal government spends 100 billion dollars on Medicare. What percent of the total budget is this?

Answer_____

9. The federal government spends 17 billion dollars on the environment. What percent of the total budget is this?

Answer_____

10. The federal government spends 60 billion dollars on education. What percent of the total budget is this?

Answer_____

11. If the federal government spent only 12% on defense, how much money would it spend?

Answer_____

12. If interest on the national debt increased to 27.6%, how much money would the federal government spend on interest?

Answer_____

Decimals and Percents Skills Inventory

Write as a decimal.

1. two hundredths   **2.** four and one fourth   **3.** three dollars   **4.** six cents

Change to a decimal.

5. $\frac{3}{4} =$ **6.** $\frac{1}{3} =$ **7.** $\frac{1}{10} =$ **8.** $4\frac{3}{5} =$

Change to a mixed number or fraction. Reduce if possible.

9. .8 =   **10.** .03 =   **11.** 1.75 =   **12.** 18.91 =

Compare. Write <, >, or = in each box.

13. .2 ☐ .200   **14.** 2.05 ☐ 2.5   **15.** .39 ☐ .30   **16.** 2.6 ☐ 6.2

Round to the nearest whole number.

17. 2.4 _____   **18.** .7 _____   **19.** 6.52 _____   **20.** 8.1 _____

Round to the nearest tenth.

21. 6.39 _____   **22.** .84 _____   **23.** .311 _____   **24.** 23.45 _____

Round to the nearest hundredth.

25. .357 _____   **26.** .036 _____   **27.** 4.298 _____   **28.** 4.3261 _____

Add, subtract, multiply, or divide. Round division answers to the nearest hundredth.

29.	30.	31.	32.	33.
$5.83 + 6.75	36.45 + 27.9	$4.00 − 1.97	37.2 − 4.065	10 − 4.32

34.	35.	36.	37.	38.
$8.67 × 5	$ .33 × 100	3.69 × 4.7	9.324 × .09	.3 × .2

39. $4 \overline{)\ \$83.80}$ **40.** $18 \overline{)\ \$8.90}$ **41.** $10 \overline{)\ \$ .80}$ **42.** $.5 \overline{)\ .12}$ **43.** $1.3 \overline{)\ 2.49}$

Write a ratio.

44. 7 runs in 5 innings **45.** 3 out of 6 doctors **46.** 3 pitches and 3 strikes

Solve each proportion.

47. $\dfrac{4}{6} = \dfrac{n}{3}$ **48.** $\dfrac{1}{3} = \dfrac{4}{n}$ **49.** $\dfrac{n}{7} = \dfrac{2}{14}$ **50.** $\dfrac{4}{n} = \dfrac{1}{2}$

Write a percent using the percent sign.

51. three percent = **52.** four and one half percent =

Change to a decimal or a percent.

53. $30\% =$ **54.** $106\% =$ **55.** $3\frac{1}{2}\% =$ **56.** $66\frac{2}{3}\% =$

57. $.09 =$ **58.** $.4 =$ **59.** $.25 =$ **60.** $4 =$

Change to a whole number or a fraction. Reduce if necessary.

61. $50\% =$ **62.** $8\% =$ **63.** $200\% =$ **64.** $3\frac{1}{4}\% =$

Change to a percent.

65. $\dfrac{1}{4} =$ **66.** $\dfrac{6}{100} =$ **67.** $\dfrac{5}{6} =$ **68.** $\dfrac{1}{5} =$

Compare. Use <, >, or = sign.

69.

    .5 ☐ 50%

70.

    90% ☐ 9

71.

    $\frac{1}{4}$ ☐ 4%

72.

    $16\frac{2}{3}\%$ ☐ $\frac{1}{6}$

Solve.

73.

What is 2% of 60?

74.

What is $66\frac{2}{3}\%$ of 90?

75.

What is 150% of 20?

76.

5 is 20% of what?

77.

30 is 300% of what number?

78.

125 is $1\frac{1}{4}\%$ of what number?

79.

15 is what percent of 80?

80.

99 is what percent of 33?

Below is a list of the problems in this Skills Inventory and the pages on which the skills are taught. If you missed any problems, turn to the pages listed and practice the skills. Then correct the problems you missed in the Skills Inventory.

Problem	Practice Page	Problem	Practice Page	Problem	Practice Page
Unit 1		*Unit 2*		53-56	82-84
1-2	13-14	29-30	33-36	57-60	86
3-4	17	31-33	39-42	61-64	90-91
5-8	11-12, 21-22	*Unit 3*		65-68	92-93
9-12	16	34-38	49-56	69-72	95-96
13-16	20, 23	39-43	59-68	*Unit 5*	
17-20	25	*Unit 4*		73-75	107-111, 115-116, 119
21-24	26	44-46	75	76-78	122-125, 127-128
25-28	27	47-50	76-78	79-80	131-133, 135
		51-52	81		

Glossary

addition (page 31) - Putting numbers together to find a total. The symbol + is used in addition.

$$\begin{array}{r} 1.43 \\ + 6.12 \\ \hline 7.55 \end{array}$$

average (page 89) - The amount you get when you divide a total by the number of items you added to get that total.

$$\left.\begin{array}{r} 3.2 \\ 4.7 \\ + 2. \\ \hline 9.9 \end{array}\right\} \text{3 items}$$

$$3)\overline{9.9} = 3.3$$

borrowing (page 32) - Taking an amount from a top digit in subtraction and adding it to the next digit to the right.

$$\begin{array}{r} {}^{114}\\ 3.\cancel{2}\cancel{4} \\ - 1.07 \\ \hline 2.17 \end{array}$$

carrying (page 32) - Taking an amount from the sum of digits with the same place value and adding it to the next column of digits to the left.

$$\begin{array}{r} {}^{1} \\ 3.29 \\ + 1.24 \\ \hline 4.53 \end{array}$$

chart (page 94) - Information arranged in rows and columns.

ITEM	NUMBER SOLD	COST
Rulers	8	$ 8.00
Scissors	2	1.68
Pencils	16	.48
Staplers	4	12.40
Pens	12	9.48
Notepads	6	2.70
TOTAL	48	$34.74

circle graph (page 150) - A circle cut into sections to show the parts that make a total.

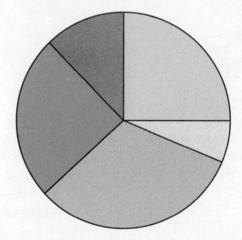

column (page 43) - A vertical line of numbers.

3 2 9
5 4 1
0 6 7

comparing (page 23) - Deciding if a number is equal to, greater than, or less than another number.

complex percent (page 84) - A percent that has a whole number and a fraction.

$33\frac{1}{3}\%$

cross-multiplying (page 76) - Multiplying the numbers on opposite corners of a proportion.

$\frac{1}{3} \diagup\!\!\!\!\diagdown \frac{n}{15}$

$3 \times n = 1 \times 15$

$3n = 15$

$n = 5$

decimal (page 9) - A number or part of a number that is less than 1.

.13   6.42

decimal point (page 11) - A decimal point separates the whole number and decimal part of a decimal.

1.2

denominator (page 10) - The bottom number in a fraction. The number of equal parts in the whole.

$\frac{2}{3}$

digit (page 13) - One of the ten symbols used to write numbers.

0 1 2 3 4 5 6 7 8 9

division (page 21) - Splitting an amount into equal groups. The symbols $\div$ and $\overline{)}$ are used in division.

$1.2 \div 2 = .6$ $2\overline{)1.2}\,^{.6}$

equal (page 20) - The same in value. The symbol = means *equal*.

6 = 6.0

equivalent (page 84) - The same in value.

$1\frac{1}{2}\% = .015\%$

estimating (page 69) - Finding an answer by rounding the numbers in a problem. You use estimating when an exact answer is not needed.

formula (page 113) - A mathematical sentence that uses letters to show a relationship.

I = prt

fraction (page 9) - Part of a whole or a group.

$\frac{3}{10}$

fraction bar (page 21) - The line that separates the numerator and denominator of a fraction.

$\frac{2}{3}$

greater than (page 23) - More than. The symbol > means *greater than*.

1 > .1 means 1 is greater than .1

higher terms (page 10) - A fraction is in higher terms when you multiply the numerator and denominator by the same number.

$\frac{1}{2} = \frac{1 \times 2}{2 \times 2} = \frac{2}{4}$

hundredth (page 14) - A decimal with two places to the right of the decimal point.

two hundredths = .02

improper fraction (page 90) - A fraction with the numerator equal to or larger than the denominator.

$\frac{4}{4}$ $\frac{7}{4}$

invert (page 124) - Turn upside down.

$\frac{1}{2} \diagup\!\!\!\!\diagdown \frac{2}{1}$

less than (page 23) - Smaller than. The symbol < means *less than*.

.1 < 1 means .1 is less than 1

lowest terms (page 10) - A fraction is in lowest terms when 1 is the only number that divides evenly into both the numerator and the denominator.

$$\frac{4}{8} = \frac{4 \div 4}{8 \div 4} = \frac{1}{2}$$

minus (page 118) - To subtract. The symbol for minus is $-$.

mixed number (page 11) - A number with a whole number part and a fraction part.

$2\frac{1}{2}$

multiplication (page 12) - Combining equal numbers two or more times to get a total. The symbol $\times$ is used in multiplication.

$$\begin{array}{r} 4 \\ 4.7 \\ \times\ \ 6 \\ \hline 28.2 \end{array}$$

n (page 77) - A symbol used to stand for a missing number.

not equal (page 20) - Different in value. The symbol $\neq$ means *not equal*.

$4.1 \neq 4.01$

numerator (page 10) - The top number in a fraction. The number of parts being considered.

$\frac{5}{9}$

operation (page 141) - The process you use to solve a math problem. The basic operations are addition, subtraction, multiplication, and division.

partial product (page 51) - The total you get when you multiply a number by one digit of another number.

$$\begin{array}{r} 2.31 \\ \times\ \ 1.2 \\ \hline 462 \\ 2\ 31 \\ \hline 2.772 \end{array}$$

percent (page 73) - A percent is a part of something. The symbol % is used to show percent.

twenty-five percent = 25%

plus (page 32) - To add. The symbol for plus is +.

proportion (page 73) - Two equal ratios or fractions.

$$\frac{3}{4} = \frac{9}{12}$$

ratio (page 73) - A fraction showing the relationship of two numbers.

3 out of 4 doctors $\frac{3}{4}$

reducing (page 10) - Dividing both the numerator and denominator of a fraction by the same number.

$$\frac{6}{9} = \frac{6 \div 3}{9 \div 3} = \frac{2}{3}$$

remainder (page 22) - the numerator of a fraction when the amount left over in a division problem is shown as a fraction.

$$\begin{array}{r} .33\ \tfrac{1}{3} \\ 3)\overline{1.00} \\ -\ \ 9 \\ \hline 10 \\ -\ \ 9 \\ \hline 1 \end{array}$$

renaming (page 32) - Carrying or borrowing a number.

$$\begin{array}{r} 1 \\ 2.6 \\ + 1.9 \\ \hline 4.5 \end{array} \qquad \begin{array}{r} 3\ 16 \\ \cancel{4}.\cancel{6} \\ - 1.9 \\ \hline 2.7 \end{array}$$

rounding (page 25) - Expressing a number to the nearest tenth, hundredth, thousandth, and so on.

row (page 51) - A horizontal line of numbers.

3 2 9
5 4 1
0 6 7

subtraction (page 31) - Taking away a certain amount from another amount to find a difference. The symbol − is used in subtraction.

$$\begin{array}{r} 7.5 \\ - 1.4 \\ \hline 6.1 \end{array}$$

table (page 43) - Information arranged in rows and columns.

ITEM	NUMBER SOLD	COST
Rulers	8	$ 8.00
Scissors	2	1.68
Pencils	16	.48
Staplers	4	12.40
Pens	12	9.48
Notepads	6	2.70
TOTAL	48	$34.74

ten thousandth (page 19) - A decimal with four places to the right of the decimal point.

eight ten thousandths = .0008

tenth (page 13) - A decimal with one place to the right of the decimal point.

four tenths = .4

thousandth (page 19) - A decimal with three places to the right of the decimal point.

forty-five thousandths = .045

whole number (page 9) - A number that shows a whole amount.

zero (page 16) - The word name for 0.

Answers & Explanations

The answer to the problem that was worked out for you in the lesson is written here in color. The next answer has an explanation written beneath it. The answers to the rest of the problems in the lesson follow in order.

Skills Inventory

Page 6

1.	.06	**2.**	2.1
3.	$2.00	**4.**	$ .04
5.	.25	**6.**	.3
7.	$.66\frac{2}{3}$	**8.**	5.2
9.	$\frac{9}{10}$	**10.**	$\frac{1}{100}$
11.	$6\frac{1}{4}$	**12.**	$12\frac{87}{100}$
13.	.1 = .100	**14.**	1.3 > 1.03
15.	.72 > .70	**16.**	5.6 < 6.5
17.	1	**18.**	1
19.	3	**20.**	7
21.	4.3	**22.**	.7
23.	.9	**24.**	17.9
25.	.27	**26.**	.02
27.	1.20	**28.**	5.48
29.	$14.13	**30.**	39.09
31.	$3.51	**32.**	94.248
33.	3.32	**34.**	$74.00
35.	$97.00	**36.**	11.89
37.	.47154	**38.**	.08

Page 7

39.	$13.47	**40.**	$ .83
41.	$ .09	**42.**	.21
43.	1.56	**44.**	$\frac{5}{6}$
45.	$\frac{1}{10}$	**46.**	$\frac{1}{1}$
47.	2	**48.**	6
49.	3	**50.**	2
51.	1%	**52.**	$2\frac{3}{10}\%$ or 2.3%
53.	.5	**54.**	2.05
55.	.0725	**56.**	$.33\frac{1}{3}$
57.	8%	**58.**	30%
59.	75%	**60.**	200%

Page 7 (continued)

61.	$\frac{1}{4}$	**62.**	$\frac{7}{100}$
63.	3	**64.**	$\frac{1}{40}$
65.	75%	**66.**	3%
67.	$16\frac{2}{3}\%$	**68.**	60%

Page 8

69.	.2 = 20%	**70.**	70% < 7
71.	$\frac{1}{2} > 2\%$	**72.**	$33\% = \frac{1}{3}$
73.	4.5	**74.**	40
75.	11	**76.**	30
77.	25	**78.**	1,000
79.	$16\frac{2}{3}\%$	**80.**	200%

Unit 1

Page 9

1. **five hundred fifty-seven**
2. **ninety-nine**
   The number 99 has 9 tens and 9 ones.
3. **eight**
4. **two thousand forty**

thousands	hundreds	tens	ones
	3	0	5
		6	2
1,	3	2	4
	5	5	7
		9	9
			8
2,	0	4	0

Page 10

5. $\frac{50}{100}$

6. $\frac{30}{100}$

   $\frac{3}{10} = \frac{3 \times 10}{10 \times 10} = \frac{30}{100}$

7. $\frac{16}{100}$

8. $\frac{75}{100}$

159

9. $\frac{40}{100}$ **10.** $\frac{65}{100}$

11. $\frac{1}{3}$

12. $\frac{1}{4}$

$$\frac{4}{16} = \frac{4 \div 4}{16 \div 4} = \frac{1}{4}$$

13. $\frac{7}{10}$ **14.** $\frac{3}{4}$

15. $\frac{1}{3}$ **16.** $\frac{1}{20}$

17. $4\frac{3}{4}$

18. $3\frac{1}{3}$

$$3\frac{3}{9} = 3\frac{3 \div 3}{9 \div 3} = 3\frac{1}{3}$$

19. $1\frac{1}{5}$ **20.** $7\frac{1}{10}$

21. $18\frac{6}{25}$ **22.** $13\frac{1}{5}$

23. $10\frac{9}{24}$

24. $2\frac{4}{16}$

$$2\frac{1}{4} = 2\frac{1 \times 4}{4 \times 4} = 2\frac{4}{16}$$

25. $32\frac{4}{10}$

Page 11

1. $\frac{4}{10} = .4$

2. $\frac{32}{100} = .32$

The denominator is the total number of parts, 100. The numerator is the number of shaded parts, 32. $\frac{32}{100}$ is written as a decimal by writing the numerator, 32, to the right of the decimal point, .32.

3. $\frac{9}{100} = .09$ **4.** $1\frac{7}{10} = 1.7$

5. .23

6. .9

Write the numerator, 9, to the right of the decimal point, .9.

7. .87 **8.** .1

9. .43 **10.** 15.91

11. 4.8

Write a decimal point to the right of the whole number, 4. Write the numerator, 8, to the right of the decimal point, 4.8.

12. 77.53 **13.** 6.7

14. 92.5

Page 12

1. .75

2. .5

Change the fraction $\frac{1}{2}$ to a fraction with 10 as the denominator, $\frac{5}{10}$. Write the numerator, 5, to the right of the decimal point, .5.

$$\frac{1}{2} = \frac{1 \times 5}{2 \times 5} = \frac{5}{10} = .5$$

3. .44 **4.** .8

5. .15 **6.** .36

7. .6 **8.** .35

9. .25 **10.** .84

11. .58 **12.** .74

13. 7.12

14. 36.28

Change the fraction $\frac{7}{25}$ to a fraction with 100 as the denominator, $\frac{28}{100}$. Write a decimal point to the right of the whole number, 36. Write the numerator, 28, to the right of the decimal point.

$$36\frac{7}{25} = 36\frac{7 \times 4}{25 \times 4} = 36\frac{28}{100} = 36.28$$

15. 5.5 **16.** 1.4

17. 25.75 **18.** 33.3

19. 104.8 **20.** 90.05

Page 13

1. $\frac{1}{10} = .1$

2. $\frac{6}{10} = .6$

Write the numerator 6 to the right of the decimal point, .6.

3. $\frac{7}{10} = .7$ **4.** $\frac{4}{10} = .4$

5. $\frac{9}{10} = .9$ **6.** $\frac{2}{10} = .2$

7. $7\frac{5}{10} = 7.5$

8. $10\frac{2}{10} = 10.2$

Write a decimal point to the right of the whole number, 10. Write the numerator, 2, to the right of the decimal point, 10.2.

9. $96\frac{6}{10} = 96.6$ **10.** $9\frac{9}{10} = 9.9$

11. $30\frac{0}{10} = 30.0$

12. $15\frac{0}{10} = 15.0$

Write a decimal point to the right of the whole number, 15. No tenths means $\frac{0}{10}$. Write the numerator, 0, to the right of the decimal point, 15.0.

13. $42\frac{9}{10} = 42.9$ **14.** $86\frac{3}{10} = 86.3$

15. $67\frac{0}{10} = 67.0$ **16.** $70\frac{5}{10} = 70.5$

17. $6\frac{0}{10} = 6.0$ **18.** $99\frac{0}{10} = 99.0$

19. $53\frac{6}{10} = 53.6$ **20.** $21\frac{2}{10} = 21.2$

21. $80\frac{4}{10} = 80.4$ **22.** $32\frac{9}{10} = 32.9$

Page 14

1. $\frac{17}{100} = .17$

2. $\frac{26}{100} = .26$

Write the numerator, 26, to the right of the decimal point, .26.

3. $\frac{11}{100} = .11$ **4.** $\frac{50}{100} = .50$

5. $\frac{10}{100} = .10$ **6.** $\frac{70}{100} = .70$

7. $\frac{3}{100} = .03$

8. $\frac{1}{100} = .01$

Write the numerator, 1, to the right of the decimal point and insert a zero, .01.

9. $\frac{8}{100} = .08$ **10.** $10\frac{12}{100} = 10.12$

11. $30\frac{1}{100} = 30.01$

Write a decimal point to the right of the whole number, 30. Write the numerator, 1, to the right of the decimal point and insert a zero, 0.01.

12. $6\frac{0}{100} = 6.00$ **13.** $99\frac{0}{100} = 99.00$

14. $53\frac{6}{100} = 53.06$ **15.** $21\frac{10}{100} = 21.10$

16. $80\frac{14}{100} = 80.14$ **17.** $32\frac{59}{100} = 32.59$

---

1. 4.00

Write a decimal point to the right of the whole number 4. No hundredths means $\frac{00}{100}$. Write the numerator 00 to the right of the decimal point, 4.00.

2. 3.50

$3\frac{1}{2} = 3\frac{1 \times 50}{2 \times 50} = 3.50$

3. 0.00

Fernando worked 0 hours.

4. 6.25

$6\frac{1}{4} = 6\frac{1 \times 25}{4 \times 25} = 6.25$

5. 8.75

$8\frac{3}{4} = 8\frac{3 \times 25}{4 \times 25} = 8.75$

6. 2.50

$2\frac{1}{2} = 2\frac{1 \times 50}{2 \times 50} = 2.50$

Sunday	7	25
Monday	4	00
Tuesday	3	50
Wednesday	0	00
Thursday	6	25
Friday	8	75
Saturday	2	50

Page 16

1. $\frac{7}{10}$

2. $\frac{3}{5}$

Write 6 as the numerator and 10 as the denominator, $\frac{6}{10}$. Reduce.

$\frac{6}{10} = \frac{6 \div 2}{10 \div 2} = \frac{3}{5}$

3. $\frac{9}{10}$ **4.** $\frac{1}{10}$

5. $1\frac{3}{10}$

6. $6\frac{1}{2}$

Write the whole number, 6. Write 5 as the numerator and 10 as the denominator, $6\frac{5}{10}$. Reduce.

$6\frac{5}{10} = 6\frac{5 \div 5}{10 \div 5} = 6\frac{1}{2}$

7. $10\frac{2}{5}$ **8.** $8\frac{7}{10}$

9. $\frac{57}{100}$ **10.** $\frac{61}{100}$

11. $\frac{23}{25}$ **12.** $\frac{3}{5}$

13. $3\frac{11}{25}$ **14.** $12\frac{99}{100}$

15. $15\frac{13}{100}$ **16.** $29\frac{1}{4}$

17. $\frac{1}{20}$ **18.** $\frac{9}{100}$

19. $\frac{1}{100}$ **20.** $\frac{7}{100}$

21. $11\frac{3}{100}$ **22.** $29\frac{3}{50}$

23. $37\frac{1}{50}$ **24.** $75\frac{1}{100}$

25. $1\frac{83}{100}$ **26.** $94\frac{19}{25}$

27. $\frac{9}{20}$ **28.** $25\frac{21}{100}$

29. $43\frac{33}{50}$ **30.** $6\frac{1}{10}$

31. $7\frac{1}{5}$ **32.** $9\frac{93}{100}$

Page 17

1. $ .72

2. $ .09
Write a dollar sign. Write 9 to the right of the decimal point. Insert a zero.

3. $ .50   **4.** $ .80

5. $1.01   **6.** $5.07

7. $10.40   **8.** $19.66

9. $47.00   **10.** $20.00

11. forty-eight cents

12. ninety-two cents
There are no dollars so the answer is written in cents.

13. thirty cents   **14.** twenty cents

15. one cent   **16.** five cents

17. four dollars and nine cents

18. eleven dollars and two cents

19. seven dollars and sixty cents

20. twenty dollars and ten cents

21. thirty-two dollars and forty-nine cents

22. seventy-five dollars

Page 18

1. $\frac{2}{10} = .2$ **2.** $\frac{26}{100} = .26$

3. $1\frac{6}{100} = 1.06$ **4.** .5

5. 1.6   **6.** 27.95

7. 2.7   **8.** .8

9. 32.7   **10.** 12.32

11. 4.87   **12.** 63.55

13. 1.09   **14.** 72.12

15. 4.9   **16.** $\frac{6}{10} = .6$

17. $\frac{6}{100} = .06$ **18.** $1\frac{0}{100} = 1.00$

19. $9\frac{1}{10} = 9.1$ **20.** $\frac{40}{100} = .40$

21. $14\frac{2}{10} = 14.2$ **22.** $\frac{7}{100}$

23. $4\frac{1}{2}$ **24.** $89\frac{99}{100}$

25. $\frac{33}{100}$ **26.** $22\frac{2}{5}$

27. $69\frac{3}{25}$ **28.** $2.00

29. $ .10   **30.** $4.06

31. forty cents

32. one hundred six dollars and twenty-two cents

Page 19

1. sixty and three tenths

2. nineteen and seven tenths
Write the whole number, nineteen. Write *and* for the decimal point. Write the decimal part, seven. Write the place name, tenths.

3. one and seven hundredths

4. one hundred and eleven hundredths

5. eighty-nine hundredths

6. two and forty-nine hundredths

7. eleven and two hundred ninety-nine thousandths

8. ninety-nine and thirty-five thousandths

9. sixteen ten thousandths

	hundreds	tens	ones	tenths	hundredths	thousandths	ten thousandths
1.		6	0	.3			
2.		1	9	.7			
3.			1	.0	7		
4.	1	0	0	.1	1		
5.				.8	9		
6.			2	.4	9		
7.		1	1	.2	9	9	
8.		9	9	.0	3	5	
9.				.0	0	1	6

Page 20

1. $6 = $6.00
2. $1.10 ≠ $1.01
3. $4.49 ≠ $40.49
   The zero in $40.49 is in the middle and cannot be dropped, so $4.49 ≠ $40.49.
4. 5.6 = 5.600
5. 17.03 ≠ 17.3
6. .8 = .80
7. 100.02 ≠ 1.2
8. 209.0 = 209
9. 9.9 ≠ 90.90
10. 12.300 = 12.3
11. .05 ≠ .50
12. .67 = .670
13. $3.90 ≠ $3.09
14. $ .79 ≠ $7.90
15. $ .99 ≠ $9.09
16. 10 = 10.00
17. 30.0 ≠ 3
18. 100 = 100.00
19. 8.600 ≠ 8.060
20. 35.09 ≠ 35.90
21. .550 = .55

Page 21

1. .2
2. .67
   Divide the numerator, 67, by the denominator 100. Add a decimal point and two zeros. Put a decimal point in the answer. Divide until there is no remainder.

$$\begin{array}{r} .67 \\ 100\overline{)67.00} \\ -\underline{600} \\ 700 \\ -\underline{700} \\ 0 \end{array}$$

3. .9
4. .75
5. .15
6. .68
7. .55
8. .8
9. .5
10. .4
11. .88
12. .16
13. .6
14. .35
15. .52
16. .37

Page 22

1. $.16\frac{2}{3}$

2. $.66\frac{2}{3}$
   Divide the numerator, 2, by the denominator, 3. Add a decimal point and 2 zeros. Put a decimal point in the answer and divide to two decimal places, .66. Write the remainder, 2, as the fraction, $\frac{2}{3}$.

$$\begin{array}{r} .66\,\frac{2}{3} \\ 3\overline{)2.00} \\ -\underline{18} \\ 20 \\ -\underline{18} \\ 2 \end{array}$$

3. $.11\frac{1}{9}$
4. $.22\frac{2}{9}$
5. $.83\frac{1}{3}$
6. $.55\frac{5}{9}$
7. $.33\frac{1}{3}$
8. $.42\frac{6}{7}$
9. $.77\frac{7}{9}$
10. $.71\frac{3}{7}$
11. $.88\frac{8}{9}$
12. $.28\frac{4}{7}$

Page 23

1. .73 < .83
2. .9 < .91
   Add a zero to .9, .90. Line up the numbers by decimal points.
   .90
   .91
   In the tenths column 9 = 9. In the hundredths column 0 < 1, so .90 < .91.
3. .40 = .4
4. 1.362 > .363
5. 25.06 < 25.6
6. 9.09 > .099
7. 3.42 > 3.4
8. 6.92 = 6.92
9. 10.1 > 10.0
10. .88 = .880
11. 1.033 < 1.33
12. 3.30 = 3.3
13. .01 < .10
14. .10 = .100
15. .11 < 1.1

Page 24

1. The net weight shows 2.25.
   Change $2\frac{1}{4}$ to a decimal.

   $2\frac{1}{4} = 2\frac{1 \times 25}{4 \times 25} = 2\frac{25}{100} = 2.25$.

2. The fraction is $\frac{9}{10}$.
   Change 9 to a fraction, $.9 = \frac{9}{10}$.

3. The label should read .75.
   Change $\frac{3}{4}$ to a decimal.

   $\frac{3}{4} = \frac{3 \times 25}{4 \times 25} = \frac{75}{100} = .75$.

4. Less
   Change $\frac{3}{4}$ to a decimal.

   $\frac{3}{4} = \frac{3 \times 25}{4 \times 25} = .75$.
   Compare .50 to .75.
   .50
   .75
   $5 < 7$, so $.50 < .75$. Joey bought less than $\frac{3}{4}$ pound.

5. The label should read 1.75.
   Change $1\frac{3}{4}$ to a decimal.

   $1\frac{3}{4} = 1\frac{3 \times 25}{4 \times 25} = 1\frac{75}{100} = 1.75$.

6. Less
   Change $1\frac{3}{4}$ to a decimal. $1\frac{3}{4} = 1.75$.
   Compare 1.65 to 1.75.
   1.65
   1.75
   $6 < 7$, so $1.65 < 1.75$. Mary bought less than $1\frac{3}{4}$ pounds.

Page 25

1. 1
2. 7
   Look at the digit in the tenths place. Since the digit is 5, add 1 to the whole number, 6. Drop the decimal point and the digit to the right, 5. 6.5 rounds to 7.
3. 3   4. 33
5. 17   6. 93
7. 11   8. 41
9. 50   10. 0
11. 1   12. 1
13. 10   14. 39
15. 100   16. 199
17. 209   18. 305
19. 17   20. 0
21. 168

Page 26

1. 1.7
2. 7.9
   Look at the digit in the hundredths place. It is 6. Since 6 is greater than 5, add 1 to the digit in the tenths place, 8. Drop the digits to the right. 7.86 rounds to 7.9.
3. 2.9   4. 2.6
5. 7.1   6. 3.2
7. 10.6   8. 40.8
9. 50.1   10. .1
11. .9   12. .6
13. 10.0   14. 39.9
15. 100.0   16. 199.7
17. 209.1   18. 999.9
19. .0   20. .0
21. 1.0

Page 27

1. 7.35
2. 6.84
   Look at the digit in the thousandths place. Since the digit is 5 add 1 to the digit in the hundredths place, 3. Drop the digits to the right. 6.835 rounds to 6.84.
3. 2.90   4. 32.01
5. 17.05   6. 93.17
7. 10.63   8. 40.80
9. 50.08   10. .03
11. .93   12. .56
13. 8.10   14. 39.50
15. 100.00   16. 199.39
17. 209.10   18. 1.00
19. 2.00   20. 2.00
21. 2.65

Page 28

1. $.49 < .50$
   Compare .49 with the standard, .50.
   Line up the digits and compare.
   .49
   .50
   $4 < 5$, so $.49 < .50$
2. $.66 > .50$ 3. $.46 < .50$
4. $.39 < .50$ 5. $.51 > .50$

6. .65 > .50 **7.** .50 = .50
8. .44 < .50 **9.** .52 > .50
10. .47 < .50 **11.** .59 > .50
12. .64 > .50

	Example	1	2	3	4	5	6	7	8	9	10	11	12
Weight	.53	.49	.66	.46	.39	.51	.65	.50	.44	.52	.47	.59	.64
+, −, =	+	−	+	−	−	+	+	=	−	+	−	+	+

Unit 1 Review, page 29

1. .5 **2.** .75
3. .6 **4.** .68
5. .9 **6.** .26
7. $\frac{2}{10}$ = .2 **8.** $\frac{2}{100}$ = .02
9. $10\frac{4}{10}$ = 10.4 **10.** $4\frac{0}{100}$ = 4.00
11. $20\frac{6}{100}$ = 20.06 **12.** $\frac{2}{25}$
13. $\frac{3}{10}$ **14.** $\frac{7}{50}$
15. $\frac{27}{100}$ **16.** $9\frac{79}{100}$
17. $10\frac{33}{50}$ **18.** $74\frac{1}{5}$
19. $6\frac{1}{2}$ **20.** $ .95
21. $1.06 **22.** $7.00
23. sixty cents
24. two hundred seven dollars and twenty-five cents
25. twenty-eight dollars
26. thirteen dollars and three cents

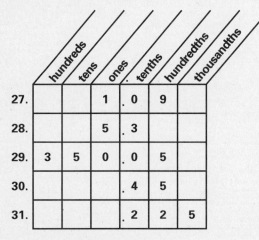

	hundreds	tens	ones	.	tenths	hundredths	thousandths
27.			1	.	0	9	
28.			5	.	3		
29.	3	5	0	.	0	5	
30.				.	4	5	
31.				.	2	2	5

27. one and nine hundredths
28. five and three tenths
29. three hundred fifty and five hundredths
30. forty-five hundredths
31. two hundred twenty-five hundredths

Page 30

32. $ .81 ≠ $8.10 **33.** 400 = 400.00
34. 6.90 ≠ 6.09 **35.** $53.00 ≠ $5.30
36. $.57\frac{1}{7}$ **37.** $.77\frac{7}{9}$
38. $.16\frac{2}{3}$ **39.** .8 = .80
40. .75 < .76 **41.** .40 < .44
42. 33.09 > 3.309 **43.** 10
44. 8 **45.** 1
46. 312 **47.** 784
48. 1.0 **49.** 19.4
50. 110.0 **51.** 302.1
52. 4.1 **53.** 1.00
54. .10 **55.** 3.16
56. 989.00 **57.** 20.10

Unit 2

Page 31

1. 484
2. 731
Subtract the ones. 8 − 7 = 1. Subtract the tens. 4 − 1 = 3. Subtract the hundredths.
9 − 2 = 7.
  948
− 217
  731
3. 64,127 **4.** 8,798
5. 15,016 **6.** 97,277

Page 32

7. 132
8. 1,430
Add the ones. 7 + 3 = 10. Carry 1 ten. Add the tens. 1 + 4 + 8 = 13. Carry 1 hundred. Add the hundreds. 1 + 1 + 2 = 4. Add the thousands. 1 + 0 = 1.
  1 1
  1,147
+   283
  1,430
9. 1,171 **10.** 80
11. 546 **12.** 3,419
13. 17,362 **14.** 163,259
15. 166

16. 381

Subtract the ones. $6 - 5 = 1$. Rename 1 hundred as 10 tens. Subtract the tens. $13 - 5 = 8$. Subtract the hundreds. $3 - 0 = 3$.

$$\begin{array}{r} {\scriptstyle 3\ 13} \\ 4\cancel{3}6 \\ -\ \ 55 \\ \hline 381 \end{array}$$

17. 463   **18.** 3
19. 689   **20.** 1,114
21. 41,199   **22.** 94,905
23. 550
24. 6,700

Add the ones. $0 + 0 = 0$. Add the tens. $0 + 0 = 0$. Add the hundreds. $7 + 0 = 7$. Add the thousands. $4 + 2 = 6$.

$$\begin{array}{r} 4,700 \\ +\ 2,000 \\ \hline 6,700 \end{array}$$

25. 25   **26.** 669
27. 22

Page 33

1. $9.06
2. $1.06

Line up the decimal points. Add. Begin with the digits on the right. $0 + 6 = 6$. $4 + 6 = 10$. Put a decimal point and a dollar sign in the answer.

$$\begin{array}{r} \$\ .40 \\ +\ \ .66 \\ \hline \$1.06 \end{array}$$

3. $2.83   **4.** $2.65
5. $11.63   **6.** $10.72
7. $8.66   **8.** $9.98
9. $18.52   **10.** $6.56
11. $7.62   **12.** $8.28

Page 34

1. $723.29
2. $51.40

Line up the decimal points. Add a decimal point and two zeros. $7 = $7.00. Add. Begin with the digits on the right. $0 + 0 = 0$. $4 + 0 = 4$. $4 + 7 = 11$. Rename. $1 + 4 = 5$.

$$\begin{array}{r} {\scriptstyle 1} \\ \$44.40 \\ +\ \ 7.00 \\ \hline \$51.40 \end{array}$$

3. $859.02   **4.** $424.99
5. $406.87

Line up the decimal points. Add a decimal point and two zeros. $340 = $340.00. Add. Begin with the digits on the right. $0 + 9 + 8 = 17$. Rename. $1 + 7 = 8$. $0 + 6 = 6$. $4 + 6 = 10$. $1 + 3 = 4$. Rename.

$$\begin{array}{r} {\scriptstyle 1 \qquad 1} \\ \$340.00 \\ 66.09 \\ +\ \ \ \ \ .78 \\ \hline \$406.87 \end{array}$$

6. $504.33   **7.** $508.75
8. $39.99

Add a decimal point and two zeros. $10 = $10.00. Add $10.00 + $29.99.

$$\begin{array}{r} \$10.00 \\ +\ 29.99 \\ \hline \$39.99 \end{array}$$

The total cost of the clothes was $39.99.

9. $37.80

Add a decimal point and two zeros. $35 = $35.00. Add $35.00 + $2.80.

$$\begin{array}{r} \$35.00 \\ +\ \ 2.80 \\ \hline \$37.80 \end{array}$$

Lee spent $37.80 on shoes.

Page 35

1. 45.6
2. 148.9

Add a decimal point and a zero. $49 = 49.0$. Add. Begin with the digits on the right. $9 + 0 = 9$. $9 + 9 = 18$. Rename. $1 + 9 + 4 = 14$.

$$\begin{array}{r} {\scriptstyle 1} \\ 99.9 \\ +\ 49.0 \\ \hline 148.9 \end{array}$$

3. 79.27   **4.** 45.6
5. 292.03   **6.** 1.111
7. 115.01   **8.** 14.2
9. 20.00   **10.** 10.010
11. 44.0   **12.** 1.29
13. 143.05   **14.** 93.731

15. 515.009   **16.**   25.33
17. 155.00

Page 36

1. .809

2. 1.9109

Add a zero. .604 = .6040. Add. Begin
with the digits on the right. 0 + 9 = 9.
4 + 6 = 10. Rename. 1 + 0 + 0 = 1.
6 + 3 = 9.   0 + 1 = 1.

      1
   .6040
+ 1.3069
  1.9109

3. 20.309   **4.**   81.4455
5. 213.737   **6.**   16.2033
7. 59.5111   **8.**   15.8228
9. 143.2055   **10.**   14.599
11. 14.925
12. 786.59

Line up the digits. Add a decimal point
and zeros. 1 = 1.00. Add a zero. .5 = .50.
Add. Begin with the digits on the right.
9 + 0 + 0 = 9. 0 + 0 + 5 = 5. 5 + 1 = 6.
8 + 0 = 8. 7 + 0 = 7.

  785.09
    1.00
+   .50
  786.59

13. 21.71   **14.**   42.9335
15. 582.5847

Page 37

1. $4.00

$2.78 rounds to $3.00
$ .89 rounds to $1.00
  $3.00
+ 1.00
  $4.00
Macy will need about $4.00.

2. $3.00

$1.79 rounds to $2.00
$ .75 rounds to $1.00
  $2.00
+ 1.00
  $3.00
Macy spent about $3.00.

3. $30.00

$10.25 rounds to $10.00
$19.95 rounds to $20.00

$10.00
+ 20.00
 $30.00
Macy will need about $30.00.

4. $15.00

$6.95 rounds to $7.00
$7.59 rounds to $8.00
  $ 7.00
+   8.00
  $15.00
Macy will need about $15.00.

Page 38

1. .5   **2.** .16
3. .25   **4.** .28
5. .12   **6.** .8
7. $11.00   **8.** $77.30
9. $205.30   **10.** $9.28
11. $114.30   **12.** $10.90
13. $21.89   **14.** $191.29
15. 11.39   **16.** 16.46
17. 43.74   **18.** 2.22
19. 9.5113   **20.** 25.763
21. 51.97   **22.** 436.33
23. 413.951   **24.** 695.558
25. 11   **26.** 84.38
27. 161.99

Page 39

1. $2.11

2. $8.36

Line up the decimal points. Subtract.
Begin with the digits on the right.
8 − 2 = 6. 8 − 5 = 3. 8 − 0 = 8.
  $8.88
−   .52
  $8.36

3. $1.76

4. $5.87

5. $ .36

Line up the decimal points. Subtract.
Begin with the digits on the right.
Rename. 12 − 6 = 6. 5 − 2 = 3.
     512
  $ .6̸2̸
−   .26
  $ .36

6. $7.48   **7.** $3.36
8. $7.84   **9.** $1.10
10. $4.45   **11.** $5.03

12. $ .47

Page 40

1. $14.33
2. $31.87
Add a decimal point and two zeros.
$38 = $38.00. Line up the decimal
points. Subtract. Begin with the digits
on the right. Rename. $10 - 3 = 7$.
$9 - 1 = 8$. $7 - 6 = 1$. $3 - 0 = 3$.

```
        9
     7 10 10
  $3 8.0 0
 −    6.1 3
  $3 1.8 7
```

3. $119.01   **4.** $66.45
5. $247.99   **6.** $208.34
7. $150.05   **8.** $37.42
9. $3,752.50
10. $18.50
Add a decimal point and two zeros.
$20 = $20.00. Subtract.

```
        9
     1 10 10
  $2 0.0 0
 −    1.5 0
  $1 8.5 0
```
Sean has $18.50 left.

11. $4.25
Add a decimal point and two zeros.
$15 = $15.00. Subtract.

```
        9
     4 10 10
  $1 5.0 0
 −  1 0.7 5
  $   4.2 5
```
Jules got $4.25 in change.

Page 41

1. 1.27
2. 3.09
Add a zero. 8.3 = 8.30. Subtract. Begin
with the digits on the right. Rename.
$10 - 1 = 9$. $2 - 2 = 0$. $8 - 5 = 3$.

```
    2 10
   8.3 0
 − 5.2 1
   3.0 9
```

3. .99   **4.** .01
5. .87   **6.** 13.49
7. 8.7   **8.** .58

9. 8.005   **10.** .371
11. 3.66   **12.** 2.992
13. 2.91   **14.** .127
15. 5.86   **16.** .956
17. .986   **18.** 3.759
19. .145   **20.** .586

Page 42

1. .699
2. 7.172
Add a zero. 1.44 = 1.440. Subtract.
Begin with the digits on the right.
$2 - 0 = 2$. Rename. $11 - 4 = 7$.
$5 - 4 = 1$. $8 - 1 = 7$.

```
    5 11
   8.6 1 2
 − 1.4 4 0
   7.1 7 2
```

3. 11.83   **4.** 85.91
5. 87.21   **6.** 78.33
7. 85.98   **8.** 92.889
9. 880.991   **10.** 786.23
11. .74   **12.** 35.81
13. 298.969 **14.** 1.994
15. 12.5
Add a decimal point and a zero.
40 = 40.0. Subtract the number of hours
Tony worked, 27.5, from the number of
hours he needs, 40.0.

```
     9
   3 10 10
   4 0.0
 − 2 7.5
   1 2.5
```
Tony needs to work 12.5 more hours.

16. 1.71
Add a decimal point and two zeros.
8 = 8.00. Subtract the number of hours
Tony worked, 6.29, from the number of
hours he needs, 8.00.

```
      9
   7 10 10
   8.0 0
 − 6.2 9
   1.7 1
```
Tony needs to work 1.71 more hours.

Page 43

1. 2,300

Find the row that shows people with an *Elementary* education. Add the number of people with an *Elementary* education in the columns *$25,000–34,999*, *$35,000–49,999*, and *$50,000 and over*.

```
  1 1 2 1
  1,138.5
    736.0
+   425.5
  2,300.0
```

2. 24,109.9

Find the row that shows people with a *College* education. Add the number of people with a *College* education in the columns *$25,000–$34,999*, *$35,000–$49,999*, and *$50,000 and over*.

```
  1 1 1 1
   5,428.1
   7,559.3
+ 11,122.5
  24,109.9
```

Page 44

3. 1,066.9

Find the row that shows people with a *High School* education. Subtract 7,879.1 in the column *Under $10,000* from 8,946 in the column *$15,000–$24,999*.

```
      1315
    8 3 5 10
  8,9 4 6.0
− 7,8 7 9.1
  1,0 6 6.9
```

4. *Under $10,000*

Find the row that shows people with an *Elementary* education. Find the greatest number in the row, 4,945 and the name of the column it is in, *Under $10,000*.

5. *$15,000–$24,999*

Find the row that shows people with a *High School* education. Find the greatest number in the row, 8,946 and the name of the column it is in, *$15,000–$24,999*.

6. 15,121.9

Find the row that shows people who earn *Under $10,000*. Add all the numbers in the column.

7. 16,595.6

```
  2 2 2
  4,945.0
  7,879.1
+ 2,297.8
 15,121.9
```

Find the row that shows people who earn *$50,000 and over*. Add all the numbers in the column.

```
   1 1
    425.5
  5,047.6
+ 11,122.5
 16,595.6
```

8. 6,923

Find the row that shows people with an *Elementary* education. Add the number of people with an *Elementary* education in the columns *Under $10,000* and *$10,000–$14,999*.

```
  1 1 1
  4,945
+ 1,978
  6,923
```

9. $12,803.5

Find the row that shows people with a *High School* education. Add the number of people with an *Elementary* education in the columns *Under $10,000* and *$10,000–$14,999*.

```
  1 1 1
  7,879.1
+ 4,924.4
 12,803.5
```

10. 4,229.3

Find the row that shows people with one or more years of *College* education. Add the number of people with a *College* education in the columns *Under $10,000* and *$10,000–$14,999*.

```
  1 1 1
  2,297.8
+ 1,931.5
  4,229.3
```

11. *$50,000 and over*
Find the row that shows people with one or more years of *College* education. Find the greatest number in the row, 11,122.5, and the name of the column it is in, *$50,000 and over*.

12. *High School*
Find the column *$25,000–$34,999*. Compare the numbers, 1,138.5, 7,304.6, and 5,428.1. Find the row with the largest number, 7,304.6.

Unit 2 Review, page 45

1. $11.44
2. $55.17
3. $138.59
4. $12.50
5. $402.94
6. $12.70
7. $19.25
8. $283.36
9. 9.20
10. 13.56
11. 57.06
12. 84.701
13. 139.433
14. 14.749
15. 95.115
16. 423.001
17. 761.411
18. 673.142
19. 3.58
20. 100.4
21. 220.29
22. 34.0556
23. 607.0121
24. 125.4388

Page 46

25. $5.13
26. $9.20
27. $6.99
28. $45.66
29. $155.06
30. $23.96
31. $4.11
32. $70.69
33. $15.62
34. 4.691
35. 1.85
36. 6.12
37. 75.877
38. 1.885
39. 7.73
40. 111.882
41. 23.99
42. 470.009
43. 103.49
44. 8.4
45. 24.77
46. 87.30
47. 49.9968
48. 5.907
49. .01146

Unit 3

Page 47

1. **2 tenths**
2. **7 hundredths**
The 7 is in the tenths column.
3. 3 hundredths
4. 6 thousandths
5. 8 thousandths
6. 0 tenths

Page 48

7. 160
8. 11,368
Starting in the column farthest to the right, multiply. Rename when necessary.

$$\begin{array}{r} 406 \\ \times\ 28 \\ \hline 3\ 248 \\ +\ 8\ 12 \\ \hline 11,368 \end{array}$$

9. 89,523
10. 757,048
11. $7\frac{1}{3}$
12. $25\frac{25}{31}$
Set up the problem. Divide. Write the remainder 25 as the fraction $\frac{25}{31}$.

$$\begin{array}{r} 25\ \frac{25}{31} \\ 31\overline{)800} \\ -\ 62 \\ \hline 180 \\ -\ 155 \\ \hline 25 \end{array}$$

13. $100\frac{1}{8}$
14. 9
15. 1
16. 1
Look at the digit in the tenths place. Since the digit is 5, add 1 to the ones column. Drop the digits to the right. .521 rounds to 1.
17. 18
18. 10
19. 0
20. 2.1
21. 5.8
Look at the digit in the hundredths place. It is 7. Since 7 is greater than 5, add 1 to the tenths place, 7. Drop the digits to the right. 5.77 rounds to 5.8.
22. 20.2
23. 1.0
24. 45.8
25. 5.03
26. .49
Look at the digit in the thousandths place. It is 9. Since 9 is greater than 5, add 1 to the hundredths place, 8. Drop the digit to the right. .489 rounds to .49.
27. 1.00
28. 14.62
29. 200.00

Page 49

1. $14.85
2. $56.94

   Set up the problem. Multiply. Rename. Put a decimal point and a dollar sign in the answer.

   $$\begin{array}{r} {\scriptstyle 2\ 5} \\ \$9.49 \\ \times\qquad 6 \\ \hline \$56.94 \end{array}$$

3. $29.34   4. $62.80
5. $13.93   6. $25.00
7. $48.18   8. $48.00
9. $5.13
10. $ .16

   Set up the problem. Multiply. Rename. Put a decimal point and a dollar sign in the answer.

   $$\begin{array}{r} {\scriptstyle 1} \\ \$ .08 \\ \times\qquad 2 \\ \hline \$ .16 \end{array}$$

11. $2.10   12. $2.97

Page 50

1. $238.00
2. $441.28

   Set up the problem. Multiply. Put a decimal point and a dollar sign in the answer.

   $$\begin{array}{r} \$13.79 \\ \times\qquad 32 \\ \hline 27\ 58 \\ +\ 413\ 7 \\ \hline \$441.28 \end{array}$$

3. $482.41   4. $599.64
5. $1,905.12   6. $5,327.40
7. $1,681.81   8. $13,475.00
9. $3,906.00

   Multiply $325.50 for one month's rent by the 12 months in a year. Put a decimal point and a dollar sign in the answer.

   $$\begin{array}{r} \$325.50 \\ \times\qquad 12 \\ \hline 651\ 00 \\ +\ 3\ 255\ 0 \\ \hline \$3,906.00 \end{array}$$

   Adela pays $3,906.00.

10. $6,876.96

   Multiply $143.27 for car payments by 48 months.

   $$\begin{array}{r} \$143.27 \\ \times\qquad 48 \\ \hline 1\ 146\ 16 \\ +\ 5\ 730\ 8 \\ \hline \$6,876.96 \end{array}$$

   Zia pays $6,876.96.

Page 51

1. $4.90
2. $2.50

   There is one zero in 10, so add one zero to $ .25. Move the decimal point one place to the right.
   $ .25 = $ .250 = $2.50

3. $9.00   4. $23.10
5. $39.00   6. $580.60
7. $557.00
8. $1,320.00

   There are two zeros in 100, so add two zeros to $13.20. Move the decimal point two places to the right.
   $13.20 = $13.2000 = $1,320.00

9. $7,609.00   10. $24,065.00
11. $65,330.00   12. $50,003.00
13. $7,920.00
14. $1,400.00

   There are three zeros in 1,000, so add three zeros to $1.40. Move the decimal point three places to the right.
   $1.40 = $1.40000 = $1,400.00

15. $4,060.00   16. $25,290.00
17. $80,300.00   18. $320,070.00
19. $4,122.70   20. $67.00
21. $37,040.00

Page 52

1. 32.4
2. 2 or 2.0

   Set up the problem. Multiply. Put a decimal point in the answer.

   $$\begin{array}{r} .5 \\ \times 4 \\ \hline 2.0 \end{array}$$

3. 57.6   4. 114.8
5. 1.6 or 1.60

6. 63.06

Set up the problem. Multiply. Count the decimal places in the problem. Put a decimal point in the answer to show two places.

$$\begin{array}{r} \overset{3}{10.51} \\ \times\quad 6 \\ \hline 63.06 \end{array}$$

7. .24   **8.** 25.02

9. 29.2   **10.** .42

11. 3.5   **12.** 60.58

13. 306.12   **14.** 69.6

15. 1.8   **16.** 30.03

Page 53

1. 58.525

2. 1.643

Set up the problem. Multiply. Put a decimal point in the answer.

$$\begin{array}{r} .053 \\ \times\quad 31 \\ \hline 53 \\ +1\,59 \\ \hline 1.643 \end{array}$$

3. 1,289.242   **4.** 2,067.265

5. 1,131.676   **6.** 125.214

7. 5,141.65   **8.** 34,620.166

9. $1.19

Multiply the tax rate, $.085, by the amount of the purchase, $14. Put a dollar sign and a decimal point in the answer.

$$\begin{array}{r} \$.085 \\ \times\quad 14 \\ \hline 340 \\ +85 \\ \hline \$1.190 \end{array}$$

$1.190 rounds to $1.19

10. $7.80

$$\begin{array}{r} .065 \\ \times\ 120 \\ \hline 1300 \\ +65 \\ \hline \$7.800 \end{array}$$ rounds to $7.80

Page 54

1. 5

2. 32

There is one zero in 10. Move the decimal point one place to the right.
$3.2 \times 10 = 3.2 = 32$

3. 156   **4.** 1,329

5. 61.7

6. 9.8

There is one zero in 10. Move the decimal point one place to the right.
$.98 \times 10 = .98 = 9.8$

7. 723.6   **8.** 2,410.5

9. 130

10. 50

There are two zeros in 100. Move the decimal point two places to the right. Add a zero.
$.5 \times 100 = .50 = 50$

11. 860   **12.** 51,070

13. 467   **14.** 158

15. 98   **16.** 42,302

17. 500

18. 19,300

There are three zeros in 1,000. Move the decimal point three places to the right. Add two zeros.
$19.3 \times 1,000 = 19.300 = 19,300$

19. 5,700   **20.** 115,200

21. 320   **22.** 8,510

23. 1,070   **24.** 266,290

25. 8.65   **26.** 120.7

27. 10,036   **28.** 3,723.31

29. 7,314   **30.** 8.02

31. 11,006   **32.** 65,907.8

Page 55

1. 10.08

2. 28.928

Set up the problem. Multiply. Count the number of decimal places in the problem, 3. Put a decimal point in the answer to show three decimal places.

$$\begin{array}{r} 9.0\,4 \leftarrow \text{two decimal places} \\ \times\ 3.2 \leftarrow \text{one decimal place} \\ \hline 1\,8\,0\,8 \\ +27\,1\,2 \\ \hline 28.9\,2\,8 \leftarrow \text{three decimal places} \end{array}$$

3. .495   **4.** .21

5. 47.736   **6.** 35.8957

7. .23553   **8.** 18.9

9. 215.54 **10.** 4.2056
11. 46.45272 **12.** 247.408

Page 56

1. .06

2. .08
Set up the problem. Multiply. Count the number of decimal places in the problem, 2. To show two decimal places in the answer, insert a zero to the right of the decimal point.

    .4 ← one decimal place
  × .2 ← one decimal place
    .08 ← two decimal places

3. .0192 **4.** .0054
5. .0413 **6.** .06666
7. .02751 **8.** .006365
9. 1.545 **10.** .6656
11. 1.1804 **12.** 2.804

Page 57

1. $6.90
Multiply the cost of a pound of apples, $ .69, by the number of pounds, 10.

   $ .69
  ×   10
  $6.90

2. $2.67
   $ .89
  ×   3
  $2.67

3. $6.50
Multiply the cost of a pound of swordfish, $13.00, by the number of pounds, .5. Round the answer to the nearest hundredth, or cent.

   $1 3.00
  ×   .5
  $6.5 00
$6.500 rounds to $6.50.

4. $4.49
   $2.9 9
  ×   1.5
   1 4 9 5
 +2 9 9
  $4.4 8 5
$4.485 rounds to $4.49.

5. $13.51
   $5.75
 ×   2.35
   28 75
  1 72 5
+ 11 50
$13.51 25
$13.5125 rounds to $13.51.

6. $ .42
   $1.69
 ×   .25
   8 45
+ 33 8
$ .42 25
$ .4225 rounds to $ .42

Page 58

1. $4.27 **2.** $ .60
3. $13.74 **4.** $ .63
5. $14.63 **6.** $7.08
7. $26.77 **8.** $89.90
9. $1.00 **10.** $5.13
11. $1.40 **12.** 1.26
13. 153.37 **14.** 7.6
15. 1.02 **16.** .8
17. 281.2 **18.** 6.82
19. .08 **20.** 202.4
21. 1.312 **22.** 5.169
23. 15.944 **24.** 14.72 or 14.720
25. 11.2147 **26.** 13.772
27. 438.786 **28.** 15.68192
29. 1.376 **30.** .04864
31. 213.025 **32.** 1.819
33. 124.674 **34.** 1.3845
35. .00459

Page 59

1. $ .84

2. $ .31
Set up the problem. Put a decimal point and a dollar sign in the answer. Divide.

    $ .31
9)$2.79
 − 2 7
   09
  − 9
    0

3. $ .09 **4.** $ .20
5. $1.23 **6.** $2.10
7. $5.79 **8.** $8.98

Page 60

1. $ .42
2. $1.42
   Set up the problem. Put a dollar sign and a decimal point in the answer. Divide to three decimal places. Ignore the remainder. Round the answer, $1.424 to the nearest hundredth, or cent, $1.42.

   $$\begin{array}{r} \$\ 1.424 \\ 15\overline{)\$21.360} \\ -\ 15 \\ \hline 6\ 3 \\ -\ 6\ 0 \\ \hline 3\ 6 \\ -\ 3\ 0 \\ \hline 6\ 0 \\ -\ 6\ 0 \\ \hline 0 \end{array}$$

   $1.424 rounds to $1.42.
3. $16.42
4. $ .54
5. $3.65
6. $4.97
7. $ .75
8. $3.06

Page 61

1. $ .57
2. $ .04
   There is one zero in 10. Move the decimal point one place to the left. Insert a zero. Round the answer to two decimal places.
   $ .40 ÷ 10 = $0.40 = $ .040

   $ .040 rounds to $ .04.
3. $1.87
4. $ .40
5. $4.18
6. $ .08
7. $ .03
8. $ .10
   There are two zeros in 100. Move the decimal point two places to the left. Insert a zero. Round the answer to two decimal places.
   $9.89 ÷ 100 = $09.89 = $ .0989

   $ .0989 rounds to $ .10.
9. $ .10
10. $ .01
11. $ .22
12. $1.27
13. $ .50

14. $3.49
    There are three zeros in 1,000. Move the decimal point three places to the left. Round the answer to two decimal places.
    $3,485.99 ÷ 1,000 = $3485.99 = $3.48599

    $3.48599 rounds to $3.49
15. $ .70
16. $ .01
17. $ .04
18. $ .52
19. $3.08
20. $ .04
21. $ .02

Page 62

1. .92
2. 1.3
   Set up the problem. Put a decimal point in the answer. Divide.

   $$\begin{array}{r} 1.3 \\ 4\overline{)5.2} \\ -\ 4 \\ \hline 1\ 2 \\ -\ 1\ 2 \\ \hline 0 \end{array}$$

3. 1.7
4. .209
5. .37
6. 20.6
7. 2.86
8. .123
9. 3.4
10. .601
11. .21
12. 2.3

Page 63

1. 6.76
2. 9.6
   Set up the problem. Put a decimal point in the answer. Divide to two decimal places. Round 9.55 to one decimal place.

   $$\begin{array}{r} 9.55 \\ 8\overline{)76.41} \\ -\ 72 \\ \hline 4\ 4 \\ -4\ 0 \\ \hline 4\ 1 \\ -\ 4\ 0 \\ \hline 1 \end{array}$$

   9.55 rounds to 9.6.
3. 14.3
4. 1.1
5. .05

6. 40.01

Set up the problem. Put a decimal point in the answer. Divide to three decimal places. Round 40.007 to two decimal places.

```
        40.007
37)1,480.290
  − 1 48
       00
     −  0
       0 2
      −  0
         29
       −  0
         290
       − 259
          31
```

40.007 rounds to 40.01.

7. .13 **8.** .07

Page 64

1. .03

2. .54

There is one zero in 10. Move the decimal point one place to the left.
5.4 ÷ 10 = 5.4 = .54

3. 2.27 **4.** .072

5. 4.536 **6.** 29.854

7. .045

8. .002

There are two zeros in 100. Move the decimal point two places to the left. Insert two zeros.
.2 ÷ 100 = 00.2 = .002

9. .116 **10.** .0086

11. .0309 **12.** 2.4741

13. .0923

14. .0004

There are three zeros in 1,000. Move the decimal point three places to the left. Insert three zeros.
.4 ÷ 1,000 = 000.4 = .0004

15. .3116 **16.** .00047

17. .00321 **18.** .08655

19. .01802 **20.** .029004

21. .0407 **22.** .100255

23. .0573 **24.** .06102

25. .39027 **26.** 76.6341

27. .004011 **28.** .75403

29. .0886 **30.** 1.37429

Page 65

1. 30

2. 12

Set up the problem. Move the decimal point one place to the right. Add one zero. Put a decimal point in the answer. Divide.

```
        1 2.
3.5)42.0
   − 35
      70
    − 70
       0
```

3. 3,310 **4.** 310

5. 80 **6.** 300

7. 100 **8.** 200

Page 66

1. 8.6

2. 7.7

Set up the problem. Move the decimal point one place to the right. Add a zero. Put the decimal point in the answer. Divide. Add two zeros. Round the answer to one decimal place.

```
         7.69
1.3)10.0 00
    − 9 1
       90
     − 78
       1 20
     − 1 17
          3
```

7.69 rounds to 7.7.

3. 2.9 **4.** 1.8

5. 66.67 **6.** 15.41

7. 2.69 **8.** 17.67

Page 67

1. 1.3

2. 18

Set up the problem. Move both decimal points one place to the right. Put a decimal point in the answer. Divide.

```
       1 8
.3)5.4
  − 3
    2 4
  − 2 4
      0
```

3. 2   **4.** 110
5. .6   **6.** 19.4
7. 10.9   **8.** .4
9. 1.6   **10.** 6.3
11. 32.2   **12.** 2.1

Page 68

1. .6
2. .4

Set up the problem. Move both decimal points one place to the right. Put a decimal point in the answer. Divide. Round the answer, .38, to the nearest tenth.

.38 rounds to .4.

$$
\begin{array}{r}
.38 \\
.9\overline{)\,.348} \\
-\,27 \\
\hline
78 \\
-\,72 \\
\hline
6
\end{array}
$$

3. 77.4   **4.** 6.5
5. 20.99   **6.** 3.52
7. 3.11   **8.** 6.29

Page 69

1. 68 miles

Round 135.5 miles to the nearest whole number, 136. The trip was on Saturday and Sunday for a total of 2 days. To find the average number of miles traveled each day, divide 136 miles by 2 days.

$$
\begin{array}{r}
68 \\
2\overline{)\,136} \\
-\,12 \\
\hline
16 \\
-\,16 \\
\hline
0
\end{array}
$$

They traveled an average of 68 miles each day.

2. 3 miles

Round 17.9 miles to 18 miles. Round 6.2 hours to 6 hours. Divide 18 miles by 6 hours.

$$
\begin{array}{r}
3 \\
6\overline{)\,18} \\
-\,18 \\
\hline
0
\end{array}
$$

They traveled an average of 3 miles per hour.

3. 16 pounds

Round 96.1 pounds to 96 pounds. Divide 96 pounds by 6 riders.

$$
\begin{array}{r}
16 \\
6\overline{)\,96} \\
-\,6 \\
\hline
36 \\
-\,36 \\
\hline
0
\end{array}
$$

Each rider carried about 16 pounds.

4. 50 miles

Round 249.82 miles to 250 miles. Round 5.2 hours to 5 hours. Divide 250 miles by 5 hours.

$$
\begin{array}{r}
50 \\
5\overline{)\,250} \\
-\,25 \\
\hline
0 \\
-\,0 \\
\hline
0
\end{array}
$$

Katie traveled about 50 miles per hour.

Page 70

5. 25 miles

Round 9.98 gallons to 10 gallons. Round 249.82 miles to 250 miles. Divide 250 miles by 10 gallons.

$$
\begin{array}{r}
25 \\
10\overline{)\,250} \\
-\,20 \\
\hline
50 \\
-\,50 \\
\hline
0
\end{array}
$$

Katie's car travels about 25 miles on a gallon of gas.

6. $47

Round $328.60 to $329. Divide $329 by 7 days.

$$
\begin{array}{r}
\$\,47 \\
7\overline{)\,\$329} \\
-\,28 \\
\hline
49 \\
-\,49 \\
\hline
0
\end{array}
$$

Katie spent an average of $47 each day.

7. 1 pound

Round 7.6 pounds to 8 pounds. Divide 8 pounds by 8 campers.

$$
\begin{array}{r}
1 \\
8\overline{)8} \\
-8 \\
\hline
0
\end{array}
$$

Each camper ate about 1 pound of chicken.

8. $3

Round $47.96 to $48. Divide $48 by 16 pounds.

$$
\begin{array}{r}
\$\,3 \\
16\overline{)\$48} \\
-48 \\
\hline
0
\end{array}
$$

The cost was $3 per pound.

9. $70

Round $560.07 to $560. Divide $560 by 8 people.

$$
\begin{array}{r}
\$\,70 \\
8\overline{)\$560} \\
-56 \\
\hline
00 \\
-0 \\
\hline
0
\end{array}
$$

Each person paid about $70.

10. $99

Round $692.90 to $693. Divide $693 by 7 days.

$$
\begin{array}{r}
\$\,99 \\
7\overline{)\$693} \\
-63 \\
\hline
63 \\
-63 \\
\hline
0
\end{array}
$$

The group spent about $99 each day for equipment rental and supplies.

Unit 3 Review, page 71

1.	$2.68	**2.**	$5.94
3.	$37.45	**4.**	$10.80
5.	$142.44	**6.**	$1,679.52
7.	$50.00	**8.**	$2,950.00
9.	4.8	**10.**	22.4
11.	.54	**12.**	76.45
13.	30.025	**14.**	11.914
15.	210.3	**16.**	340
17.	4.48	**18.**	3.3165
19.	.63	**20.**	52.521

21.	.02	**22.**	.015
23.	.27	**24.**	.184
25.	$2.30	**26.**	$1.75
27.	$ .19	**28.**	$2.76

Page 72

29.	$20.00	**30.**	$ .10
31.	$ .03	**32.**	$ .09
33.	.05	**34.**	.22
35.	$ .01	**36.**	5.31
37.	.21	**38.**	.17
39.	4.17	**40.**	0
41.	310	**42.**	77.78
43.	2.07	**44.**	59.37

Unit 4

Page 73

1. two tenths

2. four thousandths
The 4 is in the thousandths column.

3. seven hundredths **4.** six thousandths

5. zero hundredths **6.** nine tenths

Page 74

7. .5

8. .75

Divide the numerator 3 by the denominator 4 until there is no remainder.

$$
\begin{array}{r}
.75 \\
4\overline{)3.00} \\
-28 \\
\hline
20 \\
-20 \\
\hline
0
\end{array}
$$

9.	.4	**10.**	.3
11.	.05	**12.**	.6
13.	.45	**14.**	.16

15. $.33\frac{1}{3}$

16. $.83\frac{1}{3}$

$$
\begin{array}{r}
.83 \\
6\overline{)5.00} \\
-48 \\
\hline
20 \\
-18 \\
\hline
2
\end{array}
\qquad \frac{2\div2}{6\div2}=.83\frac{1}{3}
$$

17. $.44\frac{4}{9}$ **18.** $.66\frac{2}{3}$

19. $.46\frac{2}{3}$ **20.** $.27\frac{7}{9}$

21. $.18\frac{2}{11}$ **22.** $.42\frac{6}{7}$

Page 75

1. $\frac{7}{10}$

2. $\frac{2}{3}$

Write the first number, 6, as the numerator. Write the second number, 9, as the denominator. Reduce.

$\frac{6 \div 3}{9 \div 3} = \frac{2}{3}$

3. $\frac{99}{100}$ **4.** $\frac{5}{1}$

5. $\frac{2}{1}$

Write 2 as the numerator and 1 as the denominator, $\frac{2}{1}$.

6. $\frac{5}{1}$ **7.** $\frac{1}{1}$

8. $\frac{1}{1}$ **9.** $\frac{2}{3}$

10. $\frac{8}{5}$

Write the number of hours Joe worked on Friday, 8, as the numerator and the number of hours he worked on Saturday, 5, as the denominator. $\frac{8}{5}$.

11. $\frac{100}{97}$

Write the number of dinners, 200, as the numerator and the number of guests, 194, as the denominator. $\frac{200}{194}$ reduces to $\frac{100}{97}$.

Page 76

1. $\frac{3}{8} = \frac{6}{16}$

2. $\frac{4}{9} = \frac{12}{27}$

$\frac{4}{9} = \frac{4 \times 3}{9 \times 3} = \frac{12}{27}$

Check by cross-multiplying.

$\frac{4}{9} \times \frac{12}{27}$

$4 \times 27 = 9 \times 12$

$108 = 108$

3. $\frac{7}{10} = \frac{14}{20}$ **4.** $\frac{5}{6} = \frac{30}{36}$

5. $\frac{1}{2} = \frac{2}{4}$ **6.** $\frac{2}{3} = \frac{6}{9}$

7. $\frac{4}{7} = \frac{8}{14}$

8. $\frac{3}{5} = \frac{9}{15}$

9. 40 minutes

Write 2 orders in 5 minutes as the ratio, $\frac{2}{5}$. Change $\frac{2}{5}$ to an equal ratio with 16 as the numerator.

$\frac{2}{5} = \frac{16}{\square}$

$\frac{2}{5} = \frac{2 \times 8}{5 \times 8} = \frac{16}{40}$

They need 40 minutes.

10. 10 orders

$\frac{2}{3} = \frac{\square}{15}$

$\frac{2}{3} = \frac{2 \times 5}{3 \times 5} = \frac{10}{15}$

The computer can print 10 orders.

Page 77

1. n = 1

2. n = 16

Change $\frac{4}{5}$ to an equal ratio with a denominator of 20.

$\frac{4}{5} = \frac{4 \times 4}{5 \times 4} = \frac{16}{20}$

Check by cross-multiplying.

$\frac{4}{5} \times \frac{16}{20}$

$4 \times 20 = 5 \times 16$

$80 = 80$

3. n = 7   **4.** n = 6

5. n = 1   **6.** n = 6

7. n = 2   **8.** n = 5

9. 1 video

Write 3 videos for $6 as the ratio, $\frac{3}{\$6}$. Find the equal ratio with $2 as the denominator.

$\frac{3}{\$6} = \frac{n}{\$2}$

$\frac{3}{6} = \frac{3 \div 3}{6 \div 3} = \frac{1}{2}$

n = 1

He can rent 1 video.

10. $10

Write 2 days for $5 as the ratio, $\frac{2}{\$5}$. Find the equal ratio with 4 as the numerator.

$$\frac{2}{\$5} = \frac{4}{n}$$

$$\frac{2}{5} = \frac{2 \times 2}{5 \times 2} = \frac{4}{10}$$

n = $10

It costs $10.

Page 78

1. n = 2

2. n = 16

Cross-multiply. Multiply n by the number in the opposite corner, 3, and write the answer on the left side of the equal sign. Mutiply 4 by 12 and write the answer on the right side of the equal sign. Divide the number on the right side of the equal sign, 48, by the number next to n, 3. Write n without a number next to it. Check by substituting the answer 16 for n. Cross-multiply.

$$\frac{3}{4} \bowtie \frac{12}{n}$$

$3 \times n = 4 \times 12$

3n = 48

n = 48 ÷ 3 = 16

Check.

$$\frac{3}{4} \bowtie \frac{12}{16}$$

$3 \times 16 = 4 \times 12$

48 = 48

3. n = 3   **4.** n = 5

5. n = 1   **6.** n = 30

7. n = 30   **8.** n = 1

Page 79

1. $3.38

Write a proportion. Let n be the cost of 9 cucumbers. Cross-multiply. Divide the number on the opposite side of the equal sign from n by the number next to n, 2. Divide until there are three digits to the right of the decimal point. Round the answer to the nearest cent.

$$\frac{2\ cucumbers}{\$.75} \bowtie \frac{9\ cucumbers}{n}$$

$2 \times n = \$.75 \times 9$

2n = $6.75

n = $6.75 ÷ 2 = 3.375

$3.375 rounds to $3.38, so 9 cucumbers cost $3.38.

2. $1.67

$$\frac{3\ pounds}{\$1.25} = \frac{4\ pounds}{n}$$

3n = $5

n = 1.666

$1.666 rounds to $1.67, so 4 pounds of bananas cost $1.67.

3. $1.20

$$\frac{5\ pounds}{\$3.00} = \frac{2\ pounds}{n}$$

5n = $6

2 pounds of tomatoes cost $1.20.

4. $3.33

$$\frac{3\ lemons}{\$1.00} = \frac{10\ lemons}{n}$$

3n = $10

10 lemons cost $3.33.

Page 80

1. $\frac{2}{1}$

2. $\frac{3}{4}$

$$\frac{18}{24} = \frac{3}{4}$$

3. $\frac{1}{25}$

$$\frac{4}{100} = \frac{1}{25}$$

4. $\frac{3}{\$1}$ **5.** $\frac{85}{2}$

6. $\frac{1}{1}$ **7.** n = 2

$$\frac{10}{10} = \frac{1}{1}$$

8. n = 1   **9.** n = 5

10. n = 36   **11.** n = 15

12. n = 3   **13.** n = 20

14. n = 25   **15.** n = 6

16. n = 60   **17.** n = 4

18. n = 14   **19.** n = 5

20. n = 50   **21.** n = 120

22. n = 2

23. $44

Write a proportion. Cross-multiply. Divide by the number next to n.

$$\frac{3\ days}{\$132} \bowtie \frac{1\ day}{n}$$

$3 \times n = 1 \times \$132$

3n = $132

n = $132 ÷ 3 = $44

24. $120

$$\frac{5\ \text{days}}{\$300} \bowtie \frac{2\ \text{days}}{n}$$

$$5n = \$600$$
$$n = \$600 \div 5 = \$120$$

Page 81

1. 5%
2. 8%
Write 8. Write a percent sign after the number.

3. 2%   **4.** 7%
5. 16%   **6.** 24%
7. 39%   **8.** 41%
9. 60%   **10.** 50%
11. 80%   **12.** 100%

13. $4\frac{1}{2}\%$

14. $33\frac{1}{3}\%$
Write $33\frac{1}{33}$. Write a percent sign after the number.

15. 245%   **16.** 13%

17. $2\frac{3}{10}\%$ **18.** $20\frac{1}{2}\%$

19. 2%
20. 150%
There are 100 equal parts in each whole. Write the number of shaded parts. Write a percent sign after the number.

21. 36%   **22.** 63%

Page 82

1. .35
2. .19
Write the number without a percent sign. Move the decimal point two places to the left.
19% = 19 = .19

3. .27   **4.** .42
5. .1
6. .3
Write 30% without the percent sign. Move the decimal point two places to the left. Drop the zero.
30% = 30 = .3̸0 = .3

7. .7   **8.** .9
9. .06

10. .01
Write 1% without the percent sign. Move the decimal point two places to the left. Insert a zero.
1% = 01 = .01

11. .03   **12.** .09
13. .29   **14.** .04
15. .4   **16.** .72
17. .54   **18.** .02
19. .41   **20.** .66
21. .25
Write the number without a percent sign. Move the decimal point two places to the left.
25% = 25 = .25

22. .75
75% = 75 = .75

Page 83

1. 3
2. 5
Write the number without the percent sign. Move the decimal point two places to the left. Drop the two zeros at the end.
500% = 500 = 5.0̸0̸ = 5

3. 7   **4.** 2.7
5. 1.1
Write the number without the percent sign. Move the decimal point two places to the left. Drop the zero on the end.
110% = 110 = 1.10̸ = 1.1

6. 4.5   **7.** 1.07
8. 3.08
Write the number without the percent sign. Move the decimal point two places to the left.
308% = 308 = 3.08

9. 7.01   **10.** 9.25
11. 7.14   **12.** 2.83
13. 5.2   **14.** 6.24
15. 8   **16.** 1.35
17. 4.7   **18.** 2.09
19. 2.12
Write the number without the percent sign. Move the decimal point two places to the left.
212% = 212 = 2.12

20. 1.45

$145\% = 145 = 1.45$

Page 84

1. .0225

2. .031

Change the fraction to a decimal. $\frac{1}{10} = .1$. Write the number without the percent sign. Move the decimal point two places to the left.

$3\frac{1}{10}\% = 3.1\% = 03.1 = .031$

3. .065   **4.** $.07\frac{1}{3}$

5. $.10\frac{2}{3}$

$\frac{2}{3}\%$ does not have an exact decimal equivalent, so keep the fraction. Write the number without the percent sign and move the decimal point two places to the left.

$10\frac{2}{3}\% = 10\frac{2}{3} = .10\frac{2}{3}$

6. .022   **7.** .305

8. .163   **9.** .502

10. .085   **11.** .221

12. .0184   **13.** .099

14. .028   **15.** .2516

Page 85

1. Western Bank

Change the fraction in $10\frac{1}{2}\%$ to a decimal. $\frac{1}{2} = .5$. Write the number without the percent sign and move the decimal point two places to the left.

$10\frac{1}{2}\% = .105$. Write 10.25% as a decimal, .1025. Compare .105 and .1025. .1025 < .105, so 10.25% is less than $10\frac{1}{2}\%$.

$10\frac{1}{2}\% = 10.5 = .105$ at National Bank

$10.25\% = 10.25 = .1025$ at Western Bank

.1025 < .105, so $10.25\% < 10\frac{1}{2}\%$

The interest rate at Western Bank is lower.

2. Western Bank

$10\frac{3}{4}\% = 10.75 = .1075$ at National Bank

$10.5\% = 10.5 = .105$ at Western Bank

.105 < .1075, so $10.5\% < 10\frac{3}{4}\%$

The interest rate at Western Bank is lower.

3. National Bank

$8\frac{1}{4}\% = 08.25 = .0825$ at National Bank

$8.75\% = 08.75 = .0875$ at Western Bank

.0825 < .0875, so $8\frac{1}{4}\% < 8.75\%$

The interest rate at National Bank is lower.

4. Western Bank

$8\frac{1}{4}\% = 08.25 = .0825$ at National Bank

$8.125\% = 08.125 = .08125$ at Western Bank

.08125 < .0825, so $8.125\% < 8\frac{1}{4}\%$

The interest rate at Western Bank is lower.

Page 86

1. 1%

2. 7%

Move the decimal two places to the right. Add a percent sign.

$.07 = .07 = 7\%$

3. 9%   **4.** 5%

5. 10%

6. 30%

Move the decimal two places to the right. Add a zero. Add a percent sign.

$.3 = .30 = 30\%$

7. 60%   **8.** 90%

9. 25%   **10.** 75%

11. 47%   **12.** 65%

13. 86.5%   **14.** 3.1%

15. 79.9%   **16.** 20.9%

17. $33\frac{1}{3}\%$

18. $16\frac{2}{3}\%$

$\frac{2}{3}$ does not have an exact decimal equivalent so keep the fraction. Move the decimal two places to the right. Add a percent sign.

$.16\frac{2}{3} = .16\frac{2}{3} = 16\frac{2}{3}\%$

19. $83\frac{1}{3}\%$ **20.** $16\frac{2}{3}\%$

21. 99%   **22.** 8%

23. 7.4%   **24.** 50%

25. 3.3%   **26.** 70%

27. 6%   **28.** 22.5%

Page 87

1. .28

Write the number without the percent sign. Move the decimal point two places to the left.

28% = 28 = .28

2. .08

Write the number without the percent sign. Move the decimal point two places to the left. Insert a zero.

8% = 08 = .08

3. .22

22% = 22 = .22

4. .25

25% = 25 = .25

5. .05

5% = 05 = .05

6. .16

16% = 16 = .16

7. .09

9% − 09 − .09

8. .03

3% − 03 − .03

9. .10

10% = 10 = .10

10. .04

4% = 04 = .04

Page 88

1. $\frac{1}{4}$ **2.** $\frac{1}{20}$

3. $\frac{5}{3}$ **4.** $\frac{4}{1}$

5. 6   **6.** 18

7. 15   **8.** 60

9. 25%   **10.** 100%

11. $2\frac{1}{2}\%$ **12.** 30%

13. .29   **14.** .08

15. .025   **16.** .019

17. .072   **18.** .75

19. 1.2   **20.** 4.06

21. .055   **22.** $.08\frac{1}{3}$

23. $.10\frac{2}{3}$ **24.** .195

25. 80%   **26.** 5%

27. 16%   **28.** 42.5%

29. 10%   **30.** 75%

31. 8.1%   **32.** 1%

33. 50.6%   **34.** 22%

35. 60%   **36.** 94.7%

Page 89

Batting Averages	Player	Hits	Times at Bat	Average	Percent
	Ruiz, J	35	125	.280	28%
	Lee, A.	4	20	.200	20%
	Delsanto, S.	18	75	.240	24%
	Montoya, B.	3	12	.250	25%
	Shaffer, T.	7	25	.280	28%

1. .200, 20%

Divide Lee's hits, 4, by his times at bat, 20, to find Lee's average. Write the average in thousandths. Change the average to a percent by moving the decimal point two places to the right. Add a percent sign.

$\frac{4}{20} = 4 \div 20 = .2$

$$\begin{array}{r} .2 \\ 20\overline{)4.0} \\ -4\,0 \\ \hline 0 \end{array}$$

.2 = .20 = 20%

Lee's batting average is .200. The percent is 20%.

2. .240, 24%

$\frac{18}{75} = 18 \div 75 = .24$

$$\begin{array}{r} .24 \\ 75\overline{)18.00} \\ -15\,0 \\ \hline 3\,00 \\ -3\,00 \\ \hline 0 \end{array}$$

.24 = .24 = 24%

Delsanto's batting average is .240. The percent is 24%.

3. .250, 25%

$\frac{3}{12} = 3 \div 12 = .25$

.25 = .25 = 25%

Montoya's batting average is .250. The percent is 25%.

4. .280, 28%

$\frac{7}{25} = 7 \div 25 = .28$

.28 = .28 = 28%

Shaffer's batting average is .280. The percent is 28%.

Page 90

1. $\frac{3}{4}$

182

2. $\frac{4}{25}$

Write 16 over 100 without the percent sign. Reduce.

$16\% = \frac{16}{100} = \frac{16 \div 4}{100 \div 4} = \frac{4}{25}$

3. $\frac{9}{20}$ **4.** $\frac{99}{100}$

5. $\frac{3}{10}$ **6.** $\frac{1}{2}$

7. $\frac{3}{50}$ **8.** $\frac{1}{50}$

9. $\frac{1}{20}$ **10.** $\frac{1}{4}$

11. $\frac{2}{5}$ **12.** $\frac{83}{100}$

13. $\frac{3}{1}$

14. $\frac{5}{1}$

Write 500 over 100 without the percent sign. Reduce.

$500\% = \frac{5\cancel{00}}{1\cancel{00}} = \frac{5}{1}$

15. $\frac{87}{10}$ **16.** $\frac{9}{4}$

17. $\frac{23}{5}$ **18.** $\frac{493}{50}$

19. $\frac{2}{1}$ **20.** $\frac{67}{10}$

21. $\frac{3}{5}$

Write 60% over 100 without the percent sign. Reduce.

$60\% = \frac{6\cancel{0}}{10\cancel{0}} = \frac{3}{5}$

$\frac{3}{5}$ of the customers were lost.

22. $\frac{6}{5}$

$120\% = \frac{12\cancel{0}}{10\cancel{0}} = \frac{6}{5}$

Sales increased by $\frac{6}{5}$.

Page 91

1. $\frac{3}{40}$

2. $\frac{1}{8}$

Write $12\frac{1}{2}\%$ over 100 without the percent sign. Set up a division problem. Change both numbers to improper fractions. Invert and multiply.

$\frac{12\frac{1}{2}}{100} = 12\frac{1}{2} \div 100 = \frac{25}{2} \div \frac{100}{1} = \frac{\cancel{25}}{2} \times \frac{1}{\cancel{100}} = \frac{1}{8}$

3. $\frac{1}{12}$ **4.** $\frac{2}{3}$

5. $\frac{1}{6}$ **6.** $\frac{1}{3}$

7. $\frac{1}{80}$ **8.** $\frac{3}{80}$

9. $\frac{1}{32}$ **10.** $\frac{3}{32}$

Page 92

1. 50%

2. 20%

Change $\frac{3}{15}$ to a decimal by dividing the numerator, 3, by the denominator, 15. Divide until there is no remainder. Move the decimal point two places to the right. Add a zero and a percent sign.

$$15\overline{)3.0} \quad \begin{array}{r} .2 \\ \hline -3\,0 \\ \hline 0 \end{array}$$

.2 = .20 = 20%

3. 40%   **4.** 70%

5. 36%   **6.** 35%

7. 15%   **8.** 28%

9. 20%   **10.** 64%

11. 25%   **12.** 30%

13. 90%   **14.** 60%

15. 6%   **16.** 3%

Page 93

1. $66\frac{2}{3}\%$

2. $58\frac{1}{3}\%$

Change $\frac{7}{12}$ to a decimal by dividing the numerator, 7, by the denominator, 12. Divide until the answer has two places to the right of the decimal point. Write the remainder as a fraction and reduce. Move the decimal point two places to the right. Add a percent sign.

$$12\overline{)7.00} \quad \begin{array}{r} .58 \frac{4}{12} = .58\frac{1}{3} \\ \hline -6\,0 \\ \hline 1\,00 \\ -\;\;96 \\ \hline 4 \end{array}$$

$.58\frac{1}{3} = .58\frac{1}{3} = 58\frac{1}{3}\%$

3. $44\frac{4}{9}\%$ **4.** $41\frac{2}{3}\%$

5. $83\frac{1}{3}\%$ **6.** $88\frac{8}{9}\%$

7. $6\frac{2}{3}\%$ **8.** $91\frac{2}{3}\%$

Page 94

Fraction	Decimal	Percent
$\frac{1}{100}$	.01	1 %
$\frac{1}{10}$	.1	10 %
$\frac{1}{8}$	.125	$12\frac{1}{2}$ %
$\frac{1}{6}$	$.16\frac{2}{3}$	$16\frac{2}{3}$ %
$\frac{1}{5}$	.2	20 %
$\frac{1}{4}$	.25	25 %
$\frac{3}{10}$	.3	30 %
$\frac{1}{3}$	$.33\frac{1}{3}$	$33\frac{1}{3}\%$
$\frac{3}{8}$	$.37\frac{1}{2}$	$37\frac{1}{2}\%$
$\frac{2}{5}$	.4	40 %
$\frac{1}{2}$	.5	50 %
$\frac{3}{5}$	.6	60 %
$\frac{5}{8}$	$.62\frac{1}{2}$	$62\frac{1}{2}\%$
$\frac{2}{3}$	$.66\frac{2}{3}$	$66\frac{2}{3}\%$
$\frac{7}{10}$	.7	70 %
$\frac{3}{4}$	.75	75 %
$\frac{5}{6}$	$.83\frac{1}{3}$	$83\frac{1}{3}\%$
$\frac{7}{8}$	$.87\frac{1}{2}$	$87\frac{1}{2}\%$
$\frac{9}{10}$	.9	90 %
1	1.0	100 %

Page 95

1. $14\% < 21$

2. $104\% < 1.4$

3. $8\% > .026$
Change .026 to a percent and compare it to 8%.
$.026 = .026 = 2.6\%$
$8\% > 2.6\%$, so $8\% > .026$
or
Change 8% to a decimal and compare it to .026.
$8\% = 08 = .08$
$.08 > .026$, so $8\% > .026$

4. $205\% > .25$ **5.** $36\% = .36$

6. $70\% < .9$ **7.** $.06 > 2.6\%$

8. $1.37 > 14.2\%$ **9.** $.875 > 75\%$

10. $1.3 < 1.58\%$ **11.** $6.2 > 62\frac{1}{2}\%$

12. $.013 < 13\%$ **13.** $.125 = 12\frac{1}{2}\%$

14. $.31 = 31\%$ **15.** $12.3 > 1.5\%$

16. $.6 = 60\%$

Page 96

1. $\frac{2}{5} > 20\%$

2. $\frac{1}{8} > 12\%$
Change $\frac{1}{8}$ and 12% to decimals.
Compare the decimals.
$\frac{1}{8} = .125$
$12\% = .12$
$.125 > .12$, so $\frac{1}{8} > 12\%$

3. $\frac{1}{3} = .33\frac{1}{3}$ **4.** $10\% < \frac{1}{4}$

5. $95\% > \frac{3}{4}$ **6.** $1\frac{1}{2}\% < \frac{1}{2}$

7. $\frac{1}{5} < 30\%$ **8.** $41\% < \frac{7}{10}$

9. $\frac{5}{6} = 83\frac{1}{3}\%$ **10.** $25\% = \frac{1}{4}$

11. $\frac{3}{5} > 50\%$ **12.** $75\% > \frac{2}{3}$

13. $\frac{1}{6} < 20\%$ **14.** $33\frac{1}{3}\% < \frac{3}{8}$

15. $\frac{5}{6} < 85\%$ **16.** $\frac{5}{8} > 60\%$

17. $\frac{1}{4}$ off
Change $\frac{1}{4}$ to .25. Change 20% to .20.
Compare the decimals. $.25 > .20$, so
$\frac{1}{4} > 20\%$. $\frac{1}{4}$ off is the better offer.

18. $\frac{1}{3}$ off
$.25 < .33\frac{1}{3}$, so $25\% < \frac{1}{3}$. $\frac{1}{3}$ off is the better offer.

Page 97

1. 31%
Write the amount spent for taxes, $3,720 over the amount of the total budget, $12,000. Reduce to lowest terms and change the fraction to a percent.
$\frac{\$3,720}{\$12,000} = \frac{\$372 \div 12}{\$1,200 \div 12} = \frac{31}{100}$

$\frac{31}{100} = 31\%$

Creative Cleaners spent 31% of their budget on taxes.

2. 5%

$\frac{\$600}{\$12000} = \frac{1}{20} = 5\%$

Creative Cleaners spent 5% of their budget on miscellaneous expenses.

Page 98

3. 10%

$\frac{\$1200}{\$12000} = \frac{1}{10} = 10\%$

Creative Cleaners spent 10% of their budget on advertising.

4. 12%

$\frac{\$1440}{\$12000} = \frac{12}{100} = 12\%$

Creative Cleaners spent 12% of their budget on telephone expenses.

5. 15%

$\frac{\$1800}{\$12000} = \frac{3}{20} = 15\%$

Creative Cleaners spent 15% of their budget on transportation.

6. 7%

$\frac{\$840}{\$12000} = \frac{7}{100} = 7\%$

Creative Cleaners spent 7% of their budget on insurance.

7. 6%

$\frac{\$720}{\$12000} = \frac{3}{50} = 6\%$

Creative Cleaners spent 6% of their budget on a cellular phone.

8. 50%

$\frac{\$720}{\$1440} = \frac{1}{2} = 50\%$

Creative Cleaners spent 50% of their telephone expenses on a cellular phone.

9. 30%

$\frac{\$360}{\$1200} = \frac{3}{10} = 30\%$

Creative Cleaners spent 30% of their advertising money on newspaper ads.

10. 3%

$\frac{\$360}{\$12000} = \frac{3}{100} = 3\%$

Creative Cleaners spent 3% of their budget on newspaper ads.

11. 60%

$\frac{\$1080}{\$1800} = \frac{3}{5} = 60\%$

Creative Cleaners spent 60% of their transportation expenses on gasoline.

12. 9%

$\frac{\$1080}{\$12000} = \frac{9}{100} = 9\%$

Creative Cleaners spent 9% of their budget on gasoline.

Unit 4 Review, page 99

1. $\frac{4}{3}$
2. $\frac{1}{2}$
3. $\frac{3}{100}$
4. $\frac{2}{3}$
5. n = 3
6. n = 4
7. n = 7
8. n = 8
9. 3%
10. 99%
11. $6\frac{1}{3}\%$
12. 102%
13. .45
14. .80
15. .02
16. $.33\frac{1}{3}$
17. 2.4
18. .095
19. .008
20. $.66\frac{2}{3}$
21. 30%
22. 9%
23. 11.7%
24. 25%
25. 180%
26. 101%
27. 60%
28. 99%
29. $\frac{3}{10}$
30. $\frac{9}{100}$
31. $\frac{7}{25}$
32. $\frac{3}{1}$
33. $\frac{3}{4}$
34. $\frac{1}{12}$
35. $\frac{2}{3}$
36. $\frac{6}{5}$

Page 100

37. 25%
38. $66\frac{2}{3}\%$
39. 150%
40. 635%
41. 30%
42. 16%
43. 775%
44. $933\frac{1}{3}\%$
45. .01 = 1%
46. .5 = 50%
47. $.66\frac{2}{3} = 66\frac{2}{3}\%$
48. .25 = 25%
49. .75 = 75%
50. $.33\frac{1}{3} = 33\frac{1}{3}\%$
51. .7 = 70%
52. .125 = 12.5%

53. $10\% = \frac{1}{10}$ **54.** $.5 > 5\%$

55. $\frac{3}{4} > 40\%$ **56.** $26\% < 3.1$

Unit 5

Page 101

1. .15

2. .0925

Write the number without the percent sign. Change $\frac{1}{4}$ to its decimal equivalent. Move the decimal point two places to the left. Insert a zero.

$9\frac{1}{4}\% = 09.25 = .0925$

3. .07   **4.** .05

5. .2   **6.** .33

7. .083   **8.** .99

9. 1.05   **10.** .65

11. .035   **12.** .146

13. 3   **14.** .16

15. .1   **16.** 2.5

Page 102

17. $\frac{1}{10}$

18. $\frac{1}{100}$

Write the number without the percent sign. Place 1 over 100.

$1\% = \frac{1}{100}$

19. $\frac{1}{5}$ **20.** $\frac{1}{4}$

21. $\frac{1}{3}$ **22.** $\frac{2}{3}$

23. $\frac{1}{2}$ **24.** $\frac{3}{5}$

25. $\frac{2}{5}$ **26.** $\frac{3}{4}$

27. $\frac{4}{5}$ **28.** 1

29. 1.07

30. .33

Look at the digit in the thousandths place, 3. 3 < 5 so drop the digits to the right of the hundredths place. .333 rounds to .33

31. 4.63   **32.** 10.00

33. 4.37   **34.** .70

35. 12.36   **36.** 186.00

37. 20.46   **38.** 9.93

39. 340.02   **40.** 829.41

41. .1845

42. 3.6

Multiply. Count the decimal places in the problem, 2. Put the decimal point in the answer. Drop the zero.

```
   .05
 × 72
 ──────
   10
 + 3 5
 ──────
  3.60
```

43. 21   **44.** 3

45. 20   **46.** 400

47. .77   **48.** 9

Page 103

1. part = 80
whole = 200
percent = 40%

2. part = 18
whole = 60
percent = 30%

3. part = 45
whole = 50
percent = 90%

4. part = 30
whole = 30
percent = 100%

5. part = 25
whole = 500
percent = 5%

6. part = 96
whole = 1,200
percent = 8%

7. part = 150
whole = 100
percent = 150%

8. part = 100
whole = 50
percent = 200%

9. part = 100
whole = 20
percent = 500%

10. part = 10
whole = 80
percent = 12.5%

11. part = 20
whole = 60
percent = $33\frac{1}{3}\%$

12. part = 43
whole = 1,000
percent = 4.3%

13. part = $1,800
whole = $12,000
percent = 15%

14. part = $12,800
whole = $16,000
percent = 80%

Page 104

1. whole = 40
percent = 65%
part = 40 × 65%

2. whole = 19
percent = 32%
part = 19 × 32%

Write the pieces you have. Replace the words part = whole × percent with the numbers.

3. whole = 10
percent = 6%
part = 10 × 6%

4. whole = 327
percent = 1%
part = 327 × 1%

5. whole = 700   **6.** whole = 36

percent = 250%   percent = $1\frac{1}{2}$%

part = 700 × 250%   part = $36 × 1\frac{1}{2}$%

7. whole = 80

percent = 50%

part = 80 × 50%

8. whole = 9

percent = $33\frac{1}{3}$%

part = $9 × 33\frac{1}{3}$%

Write the pieces you have. Replace the words part = whole × percent with numbers.

9. whole = 86   **10.** whole = 65

percent = 125%   percent = 10%

part = 86 × 125%   part = 65 × 10%

11. whole = 500   **12.** whole = 350

percent = $33\frac{1}{3}$%   percent = 75%

part = $500 × 33\frac{1}{3}$%   part = 350 × 75%

Page 105

1. part = 3

percent = 10%

whole = 3 ÷ 10%

2. Write the pieces you have. Replace the words whole = part ÷ percent with numbers.

part = 10

percent = 25%

whole = 10 ÷ 25%

3. part = 24   **4.** part = 15

percent = 3%   percent = 5%

whole = 24 ÷ 3%   whole = 15 ÷ 5%

5. part = 33   **6.** part = 40

percent = 110%   percent = $66\frac{2}{3}$%

whole = 33 ÷ 110%   whole = $40 ÷ 66\frac{2}{3}$%

7. part = 20

percent = 25%

whole = 20 ÷ 25%

8. part = 7

percent = 10%

whole = 7 ÷ 10%

Write the pieces you have. Replace the words whole = part ÷ percent with numbers.

9. part = 400   **10.** part = 75

percent = 200%   percent = $33\frac{1}{3}$%

whole = 400 ÷ 200%   whole = $75 ÷ 33\frac{1}{3}$%

11. part = 90   **12.** part = 130

percent = 25%   percent = 160%

whole = 90 ÷ 25%   whole = 130 ÷ 160%

Page 106

1. part = 4

whole = 16

percent = 4 ÷ 16

2. part = 10

whole = 20

percent = 10 ÷ 20

Write the pieces you have. Replace the words percent = part ÷ whole with numbers.

3. part = 4   **4.** part = 1

whole = 120   whole = 100

percent = 4 ÷ 120   percent = 1 ÷ 100

5. part = 72   **6.** part = 8

whole = 36   whole = 4

percent = 72 ÷ 36   percent = 8 ÷ 4

7. part = 10

whole = 12

percent = 10 ÷ 12

8. part = 3

whole = 2

percent = 3 ÷ 2

Write the pieces you have. Replace the words percent = part ÷ whole with numbers.

9. part = 75   **10.** part = 12

whole = 25   whole = 3

percent = 75 ÷ 25   percent = 12 ÷ 3

11. part = 150   **12.** part = 15

whole = 200   whole = 25

percent = 150 ÷ 200   percent = 15 ÷ 25

Page 107

1. 2

2. 285

Write the pieces you have. Change 95% percent to a decimal. Multiply the whole, 300, by .95.

whole = 300

percent = 95% = .95

part = 300 × .95 = 285.00 = 285

3. 3   **4.** 11

5. 45   **6.** 150

7. 42

8. 36

whole = 300

percent = 12% = .12

part = 300 × .12 = 36

9. 4

10. $150.00

whole = $1,000

percent = 15% = .15

part = $1,000 × .15 = $150.00

He pays $150.00.

11. $40.00

whole = $1,000

percent = 4% = .04

part = $1,000 × .04 = $40.00

He pays $40.00.

Page 108

1. 41

2. 66

Write the pieces you have. Change $16\frac{1}{2}\%$ to the decimal .165. Replace the words part = whole × percent with numbers.

whole = 400

percent = $16\frac{1}{2}\%$ = 16.5% = 16.5 = .165

part = 400 × .165 = 66

3. 97   **4.** 486

5. 11   **6.** 73

7. 300

8. 38

whole = 500

percent = $7\frac{3}{5}\%$ = .076

part = 500 × .076 = 38

9. 3   **10.** 147

Page 109

1. 6

2. 8

Write the pieces. Change 25% to a fraction. Reduce. Multiply the whole, 32, by $\frac{1}{4}$.

whole = 32

percent = 25% = $\frac{25}{100}$ = $\frac{1}{4}$

part = $\frac{\cancel{32}^{8}}{1} \times \frac{1}{\cancel{4}_{1}} = \frac{8}{1} = 8$

3. 36   **4.** 50

5. 40

Write the pieces. Change 50% to a fraction. Reduce. Multiply the whole, 80, by $\frac{1}{2}$.

whole = 80

percent = 50% = $\frac{5\cancel{0}}{10\cancel{0}}$ = $\frac{1}{2}$

part = $\frac{\cancel{80}^{40}}{1} \times \frac{1}{\cancel{2}_{1}} = \frac{40}{1} = 40$

6. 25

7. $60

whole = $250

percent = 24% = $\frac{24}{100}$ = $\frac{6}{25}$

part = $\frac{\cancel{250}^{10}}{1} \times \frac{6}{\cancel{25}_{1}} = \frac{60}{1}$ = $60

They spent $60.

8. $150

whole = $250

percent = 60% = $\frac{6\cancel{0}}{10\cancel{0}}$ = $\frac{3}{5}$

part = $\frac{\cancel{250}^{50}}{1} \times \frac{3}{\cancel{5}_{1}} = \frac{150}{1}$ = $150

They spent $150.

Page 110

1. 60

2. 55

Write the pieces. Change $83\frac{1}{3}\%$ to a fraction. Multiply the whole by $\frac{5}{6}$.

whole = 66

percent = $83\frac{1}{3}\%$ = $\frac{5}{6}$

part = $\frac{\cancel{66}^{11}}{1} \times \frac{5}{\cancel{6}_{1}} = \frac{55}{1} = 55$

3. 30   **4.** 180

5. 7   **6.** 240

7. 66

whole = 99

percent = $66\frac{2}{3}\%$ = $\frac{2}{3}$

part = $\frac{\cancel{99}^{33}}{1} \times \frac{2}{\cancel{3}_{1}} = \frac{66}{1} = 66$

She gave 66 shots to dogs.

8. 33

whole = 99

percent = $33\frac{1}{3}\% = \frac{1}{3}$

part = $\frac{\overset{33}{\cancel{99}}}{1} \times \frac{1}{\underset{1}{\cancel{3}}} = \frac{33}{1} = 33$

She gave 33 shots to cats.

Page 111

1. 225

2. 13

Write the pieces. Change 130% to a decimal or to an improper fraction. Reduce. Multiply the whole, 10, by the percent.

whole = 10

percent = $130\% = 1.3$ or $\frac{13}{10}$

part = $10 \times 1.3 = 13$

or

part = $\frac{\overset{1}{\cancel{10}}}{1} \times \frac{13}{\underset{1}{\cancel{10}}} = \frac{13}{1} = 13$

3. 325   **4.** 375

5. 550   **6.** 3,510

7. 435   **8.** 135

Page 112

1. 24   **2.** 5

3. 12   **4.** 24

5. 10   **6.** 2

7. 25   **8.** 29

9. 23   **10.** 90

11. 116   **12.** 735

13. $80.00

whole = $400

percent = $20\% = .20 = .2$

part = $\$400 \times .2 = \$80.00$

The downpayment was $80.00.

14. 304

whole = 320

percent = $95\% = .95$

part = $320 \times .95 = 304.00 = 304$

304 students will graduate.

Page 113

1. $51.00

Write the information. Substitute the numbers for the letters in the formula. Change the percent to a decimal. Multiply. Round your answer to the nearest hundredth.

p = $600

$r = 8\frac{1}{2}\%$

t = 1

$I = \$600 \times 8\frac{1}{2}\% \times 1$

$I = \$600 \times .085 \times 1 = \$51.000$

$51.000 rounds to $51.00

2. $215.00

p = $2,000

$r = 10\frac{3}{4}\%$

t = 1

$I = \$2,000 \times 10\frac{3}{4}\% \times 1$

$I = \$2,000 \times .1075 \times 1 = \$215.0000$

$215.0000 rounds to $215.00

Page 114

3. $126.00

p = $900

r = 7%

t = 2

$I = \$900 \times .07 \times 2 = \$126.00$

4. $375.00

p = $2,500

r = 6%

$t = 2\frac{1}{2}$

$I = \$2,500 \times .06 \times 2.5 = \$375.000$

$375.000 rounds to $375.00

5. $13.00

p = $200

$r = 6\frac{1}{2}\%$

t = 1

$I = \$200 \times .065 \times 1 = \$13.000$

$13.000 rounds to $13.00

6. $22.00

p = $22

r = 4%

t = 2

$I = \$275 \times .04 \times 2 = \$22.00$

7. $145.00

$p = \$2,000$

$r = 7\frac{1}{4}\%$

$t = 1$

$I = \$2,000 \times .0725 \times 1 = \$145.0000$

$\$145.0000$ rounds to $\$145.00$

8. $50.00

$p = \$800$

$r = 6\frac{1}{4}\%$

$t = 1$

$I = \$800 \times .0625 \times 1 = \$50.00$

9. $27.00

$p = \$400$

$r = 6\frac{3}{4}\%$

$t = 1$

$I = \$400 \times .0675 \times 1 = \$27.0000$

$\$27.0000$ rounds to $\$27.00$

10. $5.00

$p = \$100$

$r = 5\%$

$t = 1$

$I = \$100 \times .05 \times 1 = \$5.00$

Page 115

1. $\frac{n}{40} = \frac{30}{100}, 12$

2. $\frac{n}{90} = \frac{10}{100}, 9$

Write a proportion. Cross-multiply to find n.

$\frac{n}{90} \bowtie \frac{10}{100}$

$100 \times n = 10 \times 90$

$100n = 900$

$n = 900 \div 100 = 9$

3. $\frac{n}{32} = \frac{25}{100}, 8$ **4.** $\frac{n}{80} = \frac{75}{100}, 60$

5. $\frac{n}{100} = \frac{85}{100}, 85$ **6.** $\frac{n}{10} = \frac{50}{100}, 5$

7. 65

Write a proportion. Cross-multiply to find n.

$\frac{n}{125} \bowtie \frac{52}{100}$

$100 \times n = 52 \times 125$

$100n = 6,500$

$n = 6,500 \div 100 = 65$

The union has 65 employees.

8. $18.00

$\frac{n}{120} \bowtie \frac{15}{100}$

$100 \times n = 15 \times 120$

$100n = 1800$

$n = 1800 \div 100 = \$18.00$

She spends $18.00 on paper products.

Page 116

1. $\frac{n}{120} = \frac{30}{100}, 36$

2. $\frac{n}{50} = \frac{50}{100}, 25$

Write a proportion. Reduce the fraction. Cross-multiply.

$\frac{n}{50} = \frac{50}{100}$

$\frac{n}{50} \bowtie \frac{1}{2}$

$2 \times n = 1 \times 50$

$2n = 50$

$n = 50 \div 2 = 25$

3. $\frac{n}{60} = \frac{25}{100}, 15$ **4.** $\frac{n}{520} = \frac{95}{100}, 494$

5. $\frac{n}{40} = \frac{5}{100}, 2$ **6.** $\frac{n}{100} = \frac{2}{100}, 2$

7. $\frac{n}{750} = \frac{10}{100}, 75$ **8.** $\frac{n}{300} = \frac{20}{100}, 60$

9. $\frac{n}{240} = \frac{15}{100}, 36$

Page 117

1. $15,750

Peter's new salary will be 100% of his old salary plus 5% of his old salary. Write a proportion. Cross-multiply.

$\frac{n}{\$15,000} \bowtie \frac{105}{100}$

$n \times 100 = \$15,000 \times 105$

$100n = \$1,575,000$

$n = \$1,575,000 \div 100 = \$15,750$

Peter will make $15,750.

2. 48

$\frac{n}{40} \bowtie \frac{120}{100}$

$100n = 120 \times 40$

$n = 4,800 \div 100 = 48$

48 people work at Lasher's.

3. 69,000

$\frac{n}{\$60,000} \bowtie \frac{115}{100}$

$100n = 6,900,000$

$n = 6,900,000 \div 100 = 69,000$

Sklar Systems produced 69,000 stereo systems.

4. 18 miles per gallon

$$\frac{n}{15} \diagup\!\!\!\!\diagdown \frac{120}{100}$$

$100n = 1800$

$n = 1800 \div 100 = 18$

The truck got 18 miles per gallon after the tune-up.

Page 118

1. $3

The new price will be 100% of the old price minus 25% of the old price. Write a proportion. Cross-multiply.

$$\frac{n}{\$4} \diagup\!\!\!\!\diagdown \frac{75}{100}$$

$100 \times n = 75 \times \$4$

$4n = \$300$

$n = \$300 \div 100 = \$3$

The new price is $3.

2. 19,400

$$\frac{n}{24,250} \diagup\!\!\!\!\diagdown \frac{80}{100}$$

$100n = 24,250 \times 80$

$100n = 1,940,000$

$n = 1,940,000 \div 100 = 19,400$

19,400 people will ride the bus.

3. 28 hours

$$\frac{n}{40} \diagup\!\!\!\!\diagdown \frac{70}{100}$$

$100 \times n = 40 \times 70$

$100n = 2,800$

$n = 2,800 \div 100 = 28$

She will work 28 hours.

4. $150

$$\frac{n}{\$200} \diagup\!\!\!\!\diagdown \frac{75}{100}$$

$100 \times n = \$200 \times 75$

$100n = \$15,000$

$n = \$15,000 \div 100 = \$150$

Her weekly pay will be $150.

Page 119

1. $\frac{n}{90} = \frac{2}{3}$, 60

2. $\frac{n}{16} = \frac{1}{8}$, 2

Write a proportion. Write $12\frac{1}{2}\%$ as $\frac{1}{8}$. Cross-multiply to find n.

$$\frac{n}{16} \diagup\!\!\!\!\diagdown \frac{1}{8}$$

$8n = 16$

$n = 16 \div 8 = 2$

3. $\frac{n}{32} = \frac{5}{8}$, 20   **4.** $\frac{n}{6} = \frac{1}{6}$, 1

5. $\frac{n}{64} = \frac{3}{8}$, 24   **6.** $\frac{n}{240} = \frac{5}{6}$, 200

7. $150

Write a proportion. Write $33\frac{1}{3}\%$ as $\frac{1}{3}$. Cross-multiply to find n.

$$\frac{n}{\$450} \diagup\!\!\!\!\diagdown \frac{1}{3}$$

$3n = \$450$

$n = \$450 \div 3 = \$150$

A senior citizen would save $150.

8. 33

Write a proportion. Write $37\frac{1}{2}\%$ as $\frac{3}{8}$. Cross-multiply to find n.

$$\frac{n}{88} \diagup\!\!\!\!\diagdown \frac{3}{8}$$

$8 \times n = 88 \times 3$

$8n = 264$

$n = 264 \div 8 = 33$

There were 33 comedies on television last week.

Page 120

1. $10 or $10.00

Use a percent sentence.

part = $40 \times 25\%$

part = $40 \times .25 = \$10.00$

or

Use a proportion.

$$\frac{n}{40} \diagup\!\!\!\!\diagdown \frac{25}{100}$$

$100 \times n = 40 \times 25$

$100n = 1,000$

$n = 1,000 \div 100 = \$10$

You will save $10.

2. $10

part = $30 \times 33\frac{1}{3}\%$

or

$$\frac{n}{\$30} = \frac{1}{3}$$

He saved $10.

3. $140

part = $350 \times 40\%$

or

$$\frac{n}{\$350} = \frac{40}{100}$$

The discount is $140.

4. $152

part = $1,520 × 10%

or

$$\frac{n}{\$1,520} = \frac{10}{100}$$

He spent $152.

5. $384

part = $1,200 × 32%

or

$$\frac{n}{\$1,200} = \frac{32}{100}$$

It costs $384 dollars less this year.

6. $36

part = $450 × 8%

or

$$\frac{n}{\$450} = \frac{8}{100}$$

He paid $36 sales tax.

Page 121

1.	8	**2.**	45
3.	37	**4.**	204
5.	2	**6.**	5
7.	42	**8.**	32
9.	15	**10.**	48
11.	30	**12.**	364

13. $17

$$\frac{n}{\$85} = \frac{20}{100}$$

The down payment was $17.

14. $6

$$\frac{n}{\$40} = \frac{15}{100}$$

She saved $6.

Page 122

1. 190

2. 25

Write the pieces. Change 200% to a decimal. Divide the part, 50, by the percent, 2.

part = 50

percent = 200% = 2.00 = 2

whole = 50 ÷ 2 = 25

3.	140	**4.**	500
5.	24	**6.**	200
7.	200	**8.**	600

Page 123

1. 96

2. 72

Write the pieces. Change $62\frac{1}{2}$% to a decimal. Divide the part, 45, by .625.

part = 45

percent = $62\frac{1}{2}$% = .625

whole = 45 ÷ .625 = 72

3.	400	**4.**	8,800
5.	500	**6.**	200
7.	32	**8.**	400
9.	200	**10.**	200

Page 124

1. 140

2. 160

Write the pieces. Change 20% to a fraction. Reduce. Divide the part, 32, by $\frac{1}{5}$.

part = 32

percent = 20% = $\frac{20}{100} = \frac{1}{5}$

whole = $\frac{32}{1} \div \frac{1}{5} = \frac{32}{1} \times \frac{5}{1} = \frac{160}{1} = 160$

3.	8	**4.**	25
5.	40	**6.**	400

7. 2,000

part = 1,500

percent = 75% = $\frac{3}{4}$

whole = $1,500 \div \frac{3}{4} = \frac{\overset{500}{\cancel{1,500}}}{1} \times \frac{4}{\underset{1}{\cancel{3}}} = \frac{2,000}{1} = 2,000$

There are 2,000 registered voters.

8. 60

part = 12

percent = 20% = $\frac{1}{5}$

whole = $12 \div \frac{1}{5} = \frac{12}{1} \times \frac{5}{1} = \frac{60}{1} = 60$

There are 60 units.

Page 125

1. 144

2. 40

Write the pieces. Change the percent, $62\frac{1}{2}\%$, to the fraction, $\frac{5}{8}$. Divide the part, 25, by the

percent, $\frac{5}{8}$.

part = 25

percent = $62\frac{1}{2}\% = \frac{5}{8}$

whole = $\frac{25}{1} \div \frac{5}{8} = \frac{\overset{5}{\cancel{25}}}{1} \times \frac{8}{\underset{1}{\cancel{5}}} = \frac{40}{1} = 40$

3.	285	**4.**	612
5.	12	**6.**	32
7.	48	**8.**	18
9.	480	**10.**	75

Page 126

1.	5	**2.**	80
3.	2.5	**4.**	27
5.	75	**6.**	30
7.	60	**8.**	20
9.	200	**10.**	200
11.	20	**12.**	102

Page 127

1. $\frac{10}{n} = \frac{40}{100}$, 25

2. $\frac{39}{n} = \frac{1}{100}$, 3,900

Write a proportion. Cross-multiply.

$\frac{39}{n} \bowtie \frac{1}{100}$

$1 \times n = 39 \times 100$

$n = 3,900$

3. $\frac{80}{n} = \frac{10}{100}$, 800   **4.** $\frac{66}{n} = \frac{120}{100}$, 55

5. $\frac{30}{n} = \frac{75}{100}$, 40   **6.** $\frac{47}{n} = \frac{5}{100}$, 940

Page 128

1. $\frac{20}{n} = \frac{2}{3}$, 30

2. $\frac{6}{n} = \frac{1}{8}$, 48

Write a proportion. Write $12\frac{1}{2}\%$ as $\frac{1}{8}$. Cross-multiply.

$\frac{6}{n} \bowtie \frac{1}{8}$

$n \times 1 = 6 \times 8$

$n = 48$

3. $\frac{45}{n} = \frac{5}{6}$, 54   **4.** $\frac{2}{n} = \frac{1}{6}$, 12

5. $\frac{87}{n} = \frac{3}{8}$, 232   **6.** $\frac{110}{n} = \frac{1}{3}$, 330

7. $\frac{48}{n} = \frac{2}{3}$, 72   **8.** $\frac{30}{n} = \frac{5}{8}$, 48

Page 129

1. $20

Write a proportion. Reduce. Cross-multiply to find the original cost.

$\frac{\$4}{n} \bowtie \frac{20}{100}$

$20 \times n = \$4 \times 100$

$n = \$20$

The original cost was $20.

2. $350

$\frac{\$35}{n} = \frac{10}{100}$

The original cost was $350.

3. $1,000

$\frac{\$150}{n} = \frac{15}{100}$

Lisa makes $1,000 a month.

4. $28

$\frac{\$7}{n} = \frac{25}{100}$

He would have paid $28 after 10 a.m.

5. $6,000

$\frac{\$600}{n} = \frac{10}{100}$

The car costs $6,000.

6. 40

$\frac{30}{n} = \frac{75}{100}$

There are 40 stores.

Page 130

1.	150	**2.**	60
3.	9	**4.**	348
5.	40	**6.**	76
7.	5	**8.**	8
9.	30	**10.**	14
11.	8	**12.**	20

Page 131

1. 50%

2. 25%

Write the pieces. Divide the part, 27, by the whole, 108. Change the answer to a percent.

part = 27

whole = 108

percent = $27 \div 108 = .25 = 25\%$

3. 20%   **4.** 10%

5. 93%   **6.** 5%

7. 85%   **8.** 40%
9. 86%   **10.** 44%

Page 132

1. 2.5% or $2\frac{1}{2}$%

2. 2.5% or $2\frac{1}{2}$%
Write the pieces. Divide the part, 10, by the whole, 400. Change the answer to a percent.
part = 10
whole = 400
percent = 10 ÷ 400 = .025 = .025 = 2.5%
= $2\frac{1}{2}$%

3. 3.8% or $3\frac{4}{5}$%   **4.** 3.75% or $3\frac{3}{4}$%

5. 1.25% or $1\frac{1}{4}$%   **6.** 8.75% or $8\frac{3}{4}$%

7. 2.4% or $2\frac{2}{5}$%   **8.** 2.5% or $2\frac{1}{2}$%

9. 37.5% or $37\frac{1}{2}$%   **10.** 62.5% or $62\frac{1}{2}$%

11. 87.5% or $87\frac{1}{2}$%   **12.** 12.6% or $12\frac{3}{5}$%

Page 133

1. $33\frac{1}{3}$%

2. $66\frac{2}{3}$%
Write the pieces. Divide the part, 14, by the whole, 21, to the hundredths place. Write the remainder as a fraction. Change the answer to a percent.
part = 14
whole = 21

percent = 14 ÷ 21 = $.66\frac{2}{3}$ = $66\frac{2}{3}$%

3. $8\frac{1}{3}$%   **4.** $66\frac{2}{3}$%

5. $16\frac{2}{3}$%   **6.** $83\frac{1}{3}$%

7. $33\frac{1}{3}$%   **8.** $66\frac{2}{3}$%

9. $83\frac{1}{3}$%   **10.** $16\frac{2}{3}$%

Page 134

1. 50%   **2.** 25%
3. 7.5% or $7\frac{1}{2}$%   **4.** 7.5% or $7\frac{1}{2}$%
5. $16\frac{2}{3}$%   **6.** $66\frac{2}{3}$%
7. 93%   **8.** 4%

9. 20%   **10.** 1%
11. 25%   **12.** 62.5% or $62\frac{1}{2}$%
13. $33\frac{1}{3}$%   **14.** $83\frac{1}{3}$%

Page 135

1. 5%
2. 12.5%
Write a proportion. Cross-multiply to find n. Add a percent sign.
$\frac{8}{64}$ ✗ $\frac{n}{100}$
64 × n = 8 × 100
64n = 800
n = 800 ÷ 64 = .125 = 12.5%

3. $33\frac{1}{3}$%   **4.** 60%
5. 70%   **6.** 15%

Page 136

1. 50%
2. 25%
Subtract the old price from the new price. Divide by the old price. Change to a percent.
$10 − $8 = $2
$2 ÷ $8 = .25 = 25%

3. 30%   **4.** 10%
5. 25%   **6.** 100%
7. 25%   **8.** $16\frac{2}{3}$%

Page 137

1. 10%
2. 25%
Subtract the new price, 45, from the old price, 50. Divide the answer by the old price. Change to a percent.
$10 − $8 = $2
$2 ÷ $8 = .25 = 25%

3. $8\frac{1}{3}$%   **4.** $16\frac{2}{3}$%
5. 20%   **6.** 20%
7. 50%   **8.** $33\frac{1}{3}$%

Unit 5 Review, page 138

1. 54   **2.** 10
3. 10   **4.** 99
5. 15   **6.** 20
7. 3   **8.** 20
9. 500   **10.** 150

11. 64   **12.** 1,000

Page 139

13. 90   **14.** 3
15. 270   **16.** 60
17. 3.8% or $3\frac{4}{5}$%   **18.** 25%

19. $16\frac{2}{3}$%   **20.** 8.75% or $8\frac{3}{4}$%
21. 50%   **22.** 10%
23. 25%   **24.** 20%

Page 140

25. 10%   **26.** $16\frac{2}{3}$%

27. $12\frac{1}{2}$%   **28.** 20%

29. 40%   **30.** $16\frac{2}{3}$%

31. 20%   **32.** 25%

33. $33\frac{1}{3}$%   **34.** 100%

Unit **6**

Page 141

1. **5.92**
2. 12.62
$$\begin{array}{r} 12.60 \\ +\ \ .02 \\ \hline 12.62 \end{array}$$
3. 220.68   **4.** 85.1
5. 864.7   **6.** 12.97
7. 30.39   **8.** 493.064

Page 142

9. 3.9
10. .0021
$$\begin{array}{r} .03 \\ \times\ .07 \\ \hline .0021 \end{array}$$
11. 1.59   **12.** 2.22

13. $.50 = \frac{1}{2}$

14. $.02 = \frac{1}{50}$
$2\% = 02 = .02$
$2\% = \frac{2}{100} = \frac{1}{50}$

15. $.25 = \frac{1}{4}$ **16.** $.33\frac{1}{3} = \frac{1}{3}$

17. $6.70 = \frac{67}{10}$ **18.** $.16\frac{2}{3} = \frac{1}{6}$

19. whole = 40
percent = 50%
part = 20
20. whole = 90
percent = $33\frac{1}{3}\% = 33\frac{1}{3} = .33\frac{1}{3} = \frac{1}{3}$
part = $\frac{\overset{30}{\cancel{90}}}{1} \times \frac{1}{\underset{1}{\cancel{3}}} = \frac{30}{1} = 30$

21. whole = 64   **22.** whole = 500
percent = 25%   percent = 20%
part = 16   part = 100

Page 143

Date	Withdrawal		Deposit		Interest Credited		Balance	
Jan. 1							356	89
Jan. 15			37	00			393	89
Jan. 31					1	64	395	53
Feb. 10	188	23					207	30
Feb. 15	75	50					131	80
Feb. 16			450	00			581	80
Feb. 27	525	77					56	03
Feb. 28					1	96	57	99
March 15			442	01			500	00
March 21	234	75					265	25

1. $395.53
$393.89 + $1.64 = $395.53
Vanessa's new balance was $395.53.
2. $131.80
$207.30 − $75.50 = $131.80
Her new balance is $131.80.
3. b, $57.99
Add $1.96 in interest to the balance of $56.03.
$56.03 + $1.96 = $57.99
Her new balance is $57.99.
4. a, $265.25
Subtract $234.75 in tuition from a balance of $500.00.
$500.00 − $234.75 = $265.25
Her new balance is $265.25.

Page 144

1. 28.4 points
$1,192 \div 42 = 28.38$
28.38 rounds to 28.4
Williams averaged 28.4 points in a game.

2. 1,049 points
$39 \times 26.9 = 1,049.1$
1,049.1 rounds to 1,049
Maulins scored 1,049 points.

3. d, 23.9
Divide the 1,003 points by 42 games.
$1,003 \div 42 = 23.88$
23.88 rounds to 23.9

4. c, 913
Multiply 23.4 points per game by 39 games.
$23.4 \times 39 = 912.6$
912.6 rounds to 913

Page 145

1. 300 grams
$1,000 \times .3 = 300$ grams
There are 300 grams in .3 kilograms.

2. 2,100 meters
$1,000 \times 2.1 = 2,100$ meters
There are 2,100 meters in 2.1 kilometers.

3. c, .5 kilometer
Divide 500 meters by 1,000 kilometers.
$500 \div 1,000 = .5$

4. c, 7.352 kilograms
Divide 7,352 grams by 1,000 kilograms.
$7,352 \div 1,000 = 7.352$

Page 146

1. $239.45
$\$235.56 \times 1.65\% = \$235.56 \times .0165 = \$3.886740$
$3.886740 rounds to $3.89
$\$235.56 + \$3.89 = \$239.45$
The ending balance will be $239.45

2. $157.76
$\$155.43 \times 1.5\% = \$155.43 \times .015 = \$2.33145$
$2.33145 rounds to $2.33
$\$155.43 + \$2.33 = \$157.76$
The ending balance was $157.76.

Page 147

3. 24.56%
$\$24.56 \div \$100.00 = .2456$
$.2456 = .2456 = 24.56\%$
24.56% of the payment went toward the finance charge.

4. 75.44%
$\$100.00 - \$24.56 = \$75.44$
$\$75.44 \div \$100.00 = .7544$
$.7544 = .7544 = 75.44\%$
75.44% of the payment went toward the unpaid balance.

5. $169.56
$\$150.56 - \$10.00 = \$140.56$
$\$140.56 + \$29.00 = \$169.56$
The ending balance will be $169.56.

6. $145.52
$\$143.02 \times 1.75\% = \$143.02 \times .0175 = \$2.502850$
$2.502850 rounds to $2.50
$\$143.02 + \$2.50 = \$145.52$
The ending balance will be $145.52.

7. 38.9%
$\$3.89 \div \$10.00 = .389$
$.389 = .389 = 38.9\%$
38.9% of the payment went toward the finance charge.

8. 61.1%
$\$10.00 - \$3.89 = \$6.11$
$\$6.11 \div \$10.00 = .611$
$.611 = .611 = 61.1\%$
The percent of the payment that went toward the balance was 61.1%.

9. $18.07
$\$120.45 \times 15\% = \$120.45 \times .15 = \$18.0675$
$18.0675 rounds to $18.07
She paid $18.07.

10. $12.60
$\$74.00 \times 10\% = \$74.00 \times .1 = \$7.40$
$\$20.00 - \$7.40 = \$12.60$
A payment of $20.00 is $12.60 greater than a payment of 10% of the ending balance.

Page 148

1. 40 miles per hour
D = 30 miles
$T = 45 \text{ minutes} = \frac{3}{4} \text{ hour} = .75$
$R = D \div T = 30 \div .75 = 40$
He will travel 40 miles per hour.

2. 270 miles
R = 60 miles per hour
T = 4 hours and 30 minutes = $4\frac{1}{2}$ hours

$4\frac{1}{2}$ hours = 4.5 hours
D = R × T = 60 × 4.5 = 270
You can travel 270 miles.

Page 149

3. 5 hours
D = 300 miles
R = 60 miles per hour
T = D ÷ R = 300 ÷ 60 = 5 hours
It will take 5 hours.

4. 400 miles per hour
D = 300 miles

T = 45 minutes = $\frac{3}{4}$ hour = .75 hour

R = D ÷ T = 300 ÷ .75 = 400
The plane will travel 400 miles per hour.

5. 5.25 hours $5\frac{1}{4}$ hours
D = 300 miles
R = 50 miles per hour
T = D ÷ R = 300 ÷ 50 = 6 hours
6 hours − .75 hours = 5.25 hours = $5\frac{3}{4}$
hours.

It will take 5.25 hours or $5\frac{1}{4}$ hours
longer to travel by bus than plane.

6. 60 miles per hour
D = 75 miles

T = 1 hour and 15 minutes = $1\frac{1}{4}$ hours
= 1.25 hours
R = D ÷ T = 75 ÷ 1.25 = 60
His usual rate of speed is 60 miles per hour.

7. 2.5 hours or $5\frac{1}{2}$ hours
R = 30 miles per hour
D = 75
T = D ÷ R

75 ÷ 30 = 2.5 hours = $2\frac{1}{2}$ hours

It took 2.5 hours or $2\frac{1}{2}$ hours to get to Houston.

8. 17.5 miles
R = 35 miles per hour

T = 30 minutes = $\frac{1}{2}$ hour = .5 hour

D = R × T = 35 × .5 = 17.5 miles

The detour was 17.5 miles long.

9. 20 miles
D = R × T
55 × 2 = 110 miles this year
65 × 2 = 130 miles next year
130 − 110 = 20 miles
Next year you will be able to travel 20 miles farther in 2 hours.

10. 60 miles
D = R × T = 30 × 2 = 60 miles
You can travel 60 miles.

Page 150

1. 240.4 billion dollars
$\frac{n}{\$1,252} = \frac{19.2}{100}$
n = $240.384
The federal government spends about 240.4 billion dollars on Social Security.

2. 288 billion dollars
$\frac{n}{\$1,252} = \frac{23}{100}$
n = $287.96
The federal government spends about 288 billion dollars on interest payments on the national debt.

Page 151

3. 167.8 billion dollars
$\frac{n}{\$1,252} = \frac{13.4}{100}$
n = $167.76
The federal government spends about 167.8 billion dollars on health, education, and the environment.

4. 335.5 billion dollars
$\frac{n}{\$1,252} = \frac{26.8}{100}$
n = $335.53
The federal government spends about 335.5 billion dollars on other payments to individuals.

5. 12.5 billion dollars
$\frac{n}{1,252} = \frac{1}{100}$
n = $12.52
The federal government spends about 12.5 billion dollars on science, space, and technology.

6. 28.8 billion dollars

$$\frac{n}{\$1,252} = \frac{2.3}{100}$$

n = $28.79

The federal government spends about 28.8 billion dollars on veterans benefits.

7. 12%

$$\frac{150}{\$1,252} = \frac{n}{100}$$

n = 11.98

The federal government spends about 12% of the total budget on welfare payments.

8. 8%

$$\frac{100}{\$1,252} = \frac{n}{100}$$

n = 7.98

The federal government spends about 8% of the total budget on Medicare.

9. 1.4%

$$\frac{17}{\$1,252} = \frac{n}{100}$$

n = 1.35

The federal government spends about 1.4% of the total budget on environment.

10. 4.8%

$$\frac{60}{\$1,252} = \frac{n}{100}$$

n = 4.79

The federal government spends about 4.8% of the total budget on education.

11. 150.2 billion dollars

$$\frac{n}{\$1,252} = \frac{12}{100}$$

n = $150.24

The federal government would spend 150.2 billion dollars on defense.

12. 345.6 billion dollars

$$\frac{n}{\$1,252} = \frac{27.6}{100}$$

n = $345.55

The federal government would spend 345.6 billion dollars on interest.

Skills Inventory

Page 152

1. .02

2. 4.25

3. $3.00

4. $ .06

5. .75

6. .33$\frac{1}{3}$

7. .1

8. 4.6

9. $\frac{4}{5}$

10. $\frac{3}{100}$

11. 1$\frac{3}{4}$

12. 18$\frac{91}{100}$

13. .2 = .200

14. 2.05 < 2.5

15. .39 > .30

16. 2.6 < 6.2

17. 2

18. 1

19. 7

20. 8

21. 6.4

22. .8

23. .3

24. 23.5

25. .36

26. .04

27. 4.30

28. 4.33

29. $12.58

30. 64.35

31. $2.03

32. 33.135

33. 5.68

34. $43.35

35. $33

36. 17.343

37. .83916

38. .06

Page 153

39. $20.95

40. $ .49

41. $ .08

42. .24

43. 1.92

44. $\frac{7}{5}$

45. $\frac{1}{2}$

46. $\frac{1}{1}$

47. 2

48. 12

49. 1

50. 8

51. 3%

52. 4$\frac{1}{2}$% or 4.5%

53. .3

54. 1.06

55. .035

56. .66$\frac{2}{3}$

57. 9%

58. 40%

59. 25%

60. 400%

61. $\frac{1}{2}$

62. $\frac{2}{25}$

63. 2

64. $\frac{13}{400}$

65. 25%

66. 6%

67. 83$\frac{1}{3}$%

68. 20%

Page 154

69. .5 = 50%

70. 90% < 9

71. $\frac{1}{4}$ > 4%

72. 16$\frac{2}{3}$% = $\frac{1}{6}$

73. 1.2 or 1$\frac{1}{5}$

74. 60

75. 30

76. 25

77. 10

78. 10,000

79. 18.75% or 18$\frac{3}{4}$%

80. 300%

KEY OPERATION WORDS

Word problems often contain clue words that help you solve the problem. These words tell you whether you need to add, subtract, multiply, or divide. The lists of words below will help you decide which operation to use when solving word problems.

Addition

add
all together
and
both
combined
in all
increase
more
plus
sum
total

Subtraction

change (money)
decrease
difference
left
less than
more than
reduce
remain or remaining
smaller, larger, farther,
    nearer, and so on

Multiplication

in all
of
multiply
product
times (as much)
total
twice
whole

Division

average
cut
divide
each
equal pieces
every
one
split

TABLE OF MEASUREMENTS

Time

60 seconds = 1 minute
60 minutes = 1 hour
24 hours = 1 day
7 days = 1 week
52 weeks = 1 year
12 months = 1 year
365 days = 1 year

Weight

16 ounces = 1 pound
2,000 pounds = 1 ton

Length

12 inches = 1 foot
36 inches = 1 yard
3 feet = 1 yard
5,280 feet = 1 mile
1,760 yards = 1 mile

Capacity

8 ounces = 1 cup
2 cups = 1 pint
4 cups = 1 quart
2 pints = 1 quart
4 quarts = 1 gallon
8 pints = 1 gallon
16 cups = 1 gallon

EQUIVALENT FRACTIONS, DECIMALS, AND PERCENTS

Fraction	Decimal	Percent
$\frac{1}{2}$	.5	50%
$\frac{2}{2} = 1$	1.0	100%

Fraction	Decimal	Percent
$\frac{1}{3}$	$.33\frac{1}{3}$	$33\frac{1}{3}\%$
$\frac{2}{3}$	$.66\frac{2}{3}$	$66\frac{2}{3}\%$
$\frac{3}{3} = 1$	1.0	100%

Fraction	Decimal	Percent
$\frac{1}{4}$	.25	25%
$\frac{2}{4} = \frac{1}{2}$	.5	50%
$\frac{3}{4}$	.75	75%
$\frac{4}{4} = 1$	1.0	100%

Fraction	Decimal	Percent
$\frac{1}{5}$	.2	20%
$\frac{2}{5}$	.4	40%
$\frac{3}{5}$	.6	60%
$\frac{4}{5}$	.8	80%
$\frac{5}{5} = 1$	1.0	100%

Fraction	Decimal	Percent
$\frac{1}{6}$	$.16\frac{2}{3}$	$16\frac{2}{3}\%$
$\frac{2}{6} = \frac{1}{3}$	$.33\frac{1}{3}$	$33\frac{1}{3}\%$
$\frac{3}{6} = \frac{1}{2}$	.5	50%
$\frac{4}{6} = \frac{2}{3}$	$.66\frac{2}{3}$	$66\frac{2}{3}\%$
$\frac{5}{6}$	$.83\frac{1}{3}$	$83\frac{1}{3}\%$
$\frac{6}{6} = 1$	1.0	100%

Fraction	Decimal	Percent
$\frac{1}{8}$	.125	$12\frac{1}{2}\%$
$\frac{2}{8} = \frac{1}{4}$	.25	25%
$\frac{3}{8}$	$.37\frac{1}{2}$	$37\frac{1}{2}\%$
$\frac{4}{8} = \frac{1}{2}$	.5	50%
$\frac{5}{8}$	$.62\frac{1}{2}$	$62\frac{1}{2}\%$
$\frac{6}{8} = \frac{3}{4}$	.75	75%
$\frac{7}{8}$	$.87\frac{1}{2}$	$87\frac{1}{2}\%$
$\frac{8}{8} = 1$	1.0	100%

Fraction	Decimal	Percent
$\frac{1}{10}$	.1	10%
$\frac{2}{10} = \frac{1}{5}$	.2	20%
$\frac{3}{10}$	.3	30%
$\frac{4}{10} = \frac{2}{5}$	.4	40%
$\frac{5}{10} = \frac{1}{2}$	.5	50%
$\frac{6}{10} = \frac{3}{5}$	.6	60%
$\frac{7}{10}$	.7	70%
$\frac{8}{10} = \frac{4}{5}$	.8	80%
$\frac{9}{10}$	.9	90%
$\frac{10}{10} = 1$	1.0	100%

Fraction	Decimal	Percent
$\frac{1}{100}$	.01	1%
1	1.0	100%

THE PERCENT TRIANGLE

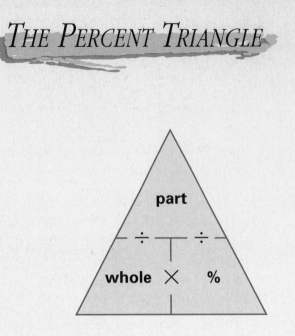

To solve for the part, cover the word *part*. The remaining pieces are connected by a multiplication sign. Multiply the pieces you have to find the part.

$$part = whole \times percent$$

To solve for the whole, cover the word *whole*. The remaining pieces are connected by a division sign. Divide the pieces you have to find the whole.

$$whole = part \div percent$$

To solve for the percent, cover the percent sign. The remaining pieces are connected by a division sign. Divide the pieces you have to find the percent.

$$percent = part \div whole$$

Addition Facts Table

+	0	1	2	3	4	5	6	7	8	9
0	0	1	2	3	4	5	6	7	8	9
1	1	2	3	4	5	6	7	8	9	10
2	2	3	4	5	6	7	8	9	10	11
3	3	4	5	6	7	8	9	10	11	12
4	4	5	6	7	8	9	10	11	12	13
5	5	6	7	8	9	10	11	12	13	14
6	6	7	8	9	10	11	12	13	14	15
7	7	8	9	10	11	12	13	14	15	16
8	8	9	10	11	12	13	14	15	16	17
9	9	10	11	12	13	14	15	16	17	18

Subtraction Facts Table

−	0	1	2	3	4	5	6	7	8	9
0	0	1	2	3	4	5	6	7	8	9
1	1	2	3	4	5	6	7	8	9	10
2	2	3	4	5	6	7	8	9	10	11
3	3	4	5	6	7	8	9	10	11	12
4	4	5	6	7	8	9	10	11	12	13
5	5	6	7	8	9	10	11	12	13	14
6	6	7	8	9	10	11	12	13	14	15
7	7	8	9	10	11	12	13	14	15	16
8	8	9	10	11	12	13	14	15	16	17
9	9	10	11	12	13	14	15	16	17	18

Multiplication Facts Table

×	0	1	2	3	4	5	6	7	8	9
0	0	0	0	0	0	0	0	0	0	0
1	0	1	2	3	4	5	6	7	8	9
2	0	2	4	6	8	10	12	14	16	18
3	0	3	6	9	12	15	18	21	24	27
4	0	4	8	12	16	20	24	28	32	36
5	0	5	10	15	20	25	30	35	40	45
6	0	6	12	18	24	30	36	42	48	54
7	0	7	14	21	28	35	42	49	56	63
8	0	8	16	24	32	40	48	56	64	72
9	0	9	18	27	36	45	54	63	72	81

Division Facts Table

÷	0	1	2	3	4	5	6	7	8	9
1	0	1	2	3	4	5	6	7	8	9
2	0	2	4	6	8	10	12	14	16	18
3	0	3	6	9	12	15	18	21	24	27
4	0	4	8	12	16	20	24	28	32	36
5	0	5	10	15	20	25	30	35	40	45
6	0	6	12	18	24	30	36	42	48	54
7	0	7	14	21	28	35	42	49	56	63
8	0	8	16	24	32	40	48	56	64	72
9	0	9	18	27	36	45	54	63	72	81

WHAT'S NEXT?

You have just finished *Decimals and Percents,* the third book in the Steck-Vaughn series, *Math Matters for Adults.*

The next book, *Measurement, Geometry, and Algebra* provides easy-to-follow steps and practice in working with measurement, geometry, and algebra. The book begins with a Skills Inventory that you can use to discover your strengths and weaknesses, and it ends with a second Skills Inventory that you can use to measure the progress you've made.

In *Measurement, Geometry, and Algebra,* you will learn to solve real-life problems using measurement, geometry, and algebra. Everyday jobs such as measuring for carpet, painting a house, or putting up a new fence involve using geometric formulas. Other questions about temperature or bank accounts can be answered using algebraic expressions or simple equations.

How many other situations can you think of in which you might use algebra and geometry? List them on the lines below.

_____

_____

_____